NOTES ON THE QUANTUM THEORY OF ANGULAR MOMENTUM

EUGENE FEENBERG
AND
GEORGE EDWARD PAKE

DOVER PUBLICATIONS, INC.
Mineola, New York

Published in Canada by General Publishing Company, Ltd., 30 Lesmill Road, Don Mills, Toronto, Ontario.

Bibliographical Note

This Dover edition, first published in 1999, is an unabridged republication of the work originally published by the Stanford University Press, Stanford, California in 1959.

Library of Congress Cataloging-in-Publication Data

Feenberg, Eugene.
Notes on the quantum theory of angular momentum / Eugene Feenberg and George Edward Pake. — Dover ed.
        p.   cm.
Originally published: Stanford, Calif. : Stanford University Press, 1959.
Includes bibliographical references and index.
ISBN 0-486-40923-6 (pbk.)
1. Angular momentum (Nuclear physics) I. Pake, G. E. (George Edward) II. Title.
QC174.17.A53F44   1999
530.12—dc21   99-35679
                                                                         CIP

Manufactured in the United States of America
Dover Publications, Inc., 31 East 2nd Street, Mineola, N.Y. 11501

PREFACE

These notes were prepared originally to assist graduate students at Washington University in reading research papers on atomic, molecular, and nuclear structure. Chapter 1 contains a review of the elements of quantum theory. The fundamental commutation relations for angular momentum components and vector operators are developed in Chapter 2. In Chapter 3 the matrix elements and eigenvalues of the angular momentum operators are worked out from the commutation relations. These chapters constitute a review on an elementary level of material usually included in a one-year course in quantum theory. The matrix elements of scalar, vector, and tensor operators are computed in Chapter 4 and applied in Chapter 5 to derive several useful relations in the theory of magnetic moments, electric quadruple moments, and dipole transition probabilities. Chapter 5 concludes with a list of references for further study.

E. F.

February 1953 G. E. P.

CONTENTS

CHAPTER 1

REVIEW OF QUANTUM THEORY DEFINITIONS AND NOTATIONS

It is assumed that the reader has encountered the material of this chapter in his previous study of quantum mechanics. If a conventional graduate course in quantum mechanics is a recent experience, the chapter need only be skimmed to gain familiarity with the notation.

1–1 Operators. An operator is a symbolic representation of a rule relating two sets of functions. Familiar examples of such rules include raising to a power, taking a logarithm, and differentiating. In quantum theory, it is customary to write the operator to the left of the function upon which it acts. Furthermore, many important physical operators are linear, a property expressed by the equation

$$A(af + bg) = a(Af) + b(Ag), \tag{1}$$

in which A is the operator, a and b are constants, and f and g are functions to which the operator is applied. Of the three operators listed above, only the derivative has the linear property.

In quantum mechanics any classical quantity expressed as a function of the particle coordinates and momenta is interpretable as an operator. The formulation in terms of the rectangular position and momentum coordinates usually allows the unambiguous identification of the operator corresponding to the classical quantity. For example, in the Schroedinger formulation of quantum mechanics, the rectangular coordinate component q_k is treated as a continuous independent variable, and the canonically conjugate momentum p_k is taken to be the derivative operator $\dfrac{\hbar}{i}\dfrac{\partial}{\partial q_k}$ ($\hbar$ is the Dirac quantum of action, Planck's h divided by 2π, and i is the imaginary unit). Thus a classical quantity $f(q_1, \ldots; p_1, \ldots)$ becomes the operator

$$f\left(q_1 \ldots, \frac{\hbar}{i}\frac{\partial}{\partial q_1}, \ldots\right),$$

1

subject to a suitable symmetrization of all factors $q_k^m p_k^n$ involving two canonically conjugate variables. It should be explicitly noted that not all operators occurring in quantum mechanics have classical counterparts. Important examples in this category are the intrinsic spin operators and the parity operator.

1-2 Linear operator algebra and matrix elements. The algebra of linear operators is similar to ordinary algebra, with one notable exception. Let A and B represent linear operators defined with respect to a function space F. Then if ψ is a function belonging to F, A and B obey the distributive law

$$(A + B)\psi = A\psi + B\psi \tag{1}$$

and the associative law

$$(AB)\psi = A(B\psi). \tag{2}$$

The second relation requires, of course, that $B\psi$ belong to F.

The notable departure from ordinary algebra lies in the possible failure of the commutative law. For certain operators A and B,

$$AB\psi \neq BA\psi \tag{3}$$

or simply $AB \neq BA$. This property is sufficiently important to warrant definition of a quantity called the *commutator* of two operators.

The commutator of A with B, written $[A,B]$, is defined as

$$[A,B] = AB - BA. \tag{4}$$

For example, the commutator of $\partial/\partial x$ with x does not vanish, and is, in fact, equal to unity:

$$\left(\frac{\partial}{\partial x} x - x \frac{\partial}{\partial x}\right)\psi = \frac{\partial}{\partial x}(x\psi) - x\frac{\partial}{\partial x}\psi = \psi. \tag{5}$$

Two operators with a vanishing commutator are said to *commute*. All multiplicative operators, such as x and y^2, commute.

The *matrix element* of an operator A with respect to the functions ϕ and ψ, written $(\phi, A\psi)$, is defined as

$$(\phi, A\psi) = \int \cdots \int \bar{\phi}(x_1, \ldots) A\psi(x_1, \ldots)\, dx_1 \ldots, \tag{6}$$

where $\bar{\phi}$ denotes the complex conjugate of ϕ. In general, the symbol (f,g), where f and g are functions of the same variables $x_1, \ldots,$ is defined as

$$(f,g) = \int \cdots \int \bar{f}(x_1, \ldots)g(x_1, \ldots)dx_1 \ldots \tag{7}$$

and is sometimes referred to as the *scalar product* of f and g.

The *adjoint* A^* of an operator A is defined by the condition

$$\begin{aligned}(\phi,A\psi) &= (A^*\phi,\psi) \\ &= \overline{(\psi,A^*\phi)}.\end{aligned} \tag{8}$$

The second equality of (8) follows directly from the definition (7) of the scalar product. We leave as exercises for the reader the proofs that A^* is a linear operator, that $(AB)^* = B^*A^*$, and that $(A^*)^* = A$.

If Eq. (8) is valid when A^* is replaced by A, then the operator A is said to be self-adjoint or *Hermitian:*

$$\begin{aligned}(\phi,A\psi) &= (A\phi,\psi) \\ &= \overline{(\psi,A\phi)}\end{aligned} \quad \begin{cases} \text{Hermitian} \\ \text{property} \end{cases}. \tag{9}$$

An important corollary of Eq. (9) is that the diagonal matrix elements (those for which $\phi = \psi$) of Hermitian operators are real. The proof is evident from Eq. (8) with ϕ replaced by ψ. In quantum mechanics, the diagonal matrix elements of an operator representing a physical quantity are interpreted as average values of the physical quantity, and the necessity that these average values be real numbers requires the operator to be Hermitian.

If A is Hermitian, the operator A^2 possesses a property with important physical consequences, namely, the diagonal matrix elements of A^2 are real, non-negative numbers:

$$\begin{aligned}(\psi,A^2\psi) &= (A^*\psi,A\psi) \\ &= (A\psi,A\psi) \geqq 0.\end{aligned} \tag{10}$$

1–3 Eigenfunctions, eigenvalues, and orthonormal sets. The possible precise values of the physical quantity corresponding to the Hermitian operator A are the numbers A_k determined by the eigenvalue problem

$$\left.\begin{array}{ll} \text{(i)} & Au_k = A_k u_k, \\ \text{(ii)} & \text{suitable boundary and} \\ & \text{continuity conditions on } u_k. \end{array}\right\} \tag{1}$$

The function u_k is called an *eigenfunction* of the operator A belonging to the eigenvalue A_k of that operator. Many problems yield both discrete and continuous ranges for the values of A_k. We consider only the discrete spectrum in order to avoid elaborate definitions and discussions out of place in this brief outline and review. Actually, the necessity for treating continuous spectra can be sidestepped by considering the physical system to be contained in an arbitrarily large box with impenetrable walls (implying $u_k = 0$ on the surface of the box).

An eigenvalue is said to be n-fold degenerate if there are n linearly independent functions $u_{k1}, u_{k2}, \ldots, u_{kn}$ belonging to the eigenvalue. It is always possible to choose these functions so that they make up an orthonormal set:

$$(u_{ki}, u_{kj}) = \int \cdots \int \overline{u_{ki}}\, u_{kj}\, d\tau = \delta_{ij}. \tag{2}$$

Then the more general orthogonality condition

$$(u_{ki}, u_{lj}) = \int \cdots \int \overline{u_{ki}}\, u_{lj}\, d\tau = \delta_{kl}\delta_{ij} \tag{3}$$

is a consequence of Eq. (1) and the Hermitian property of A. Equation (3) characterizes an orthonormal set of eigenfunctions of A.

1–4 Commuting operators and simultaneous eigenfunctions. If A and B commute, the eigenfunctions of the two operators are closely related. The application of B to the eigenvalue equation

$$Au_{ki} = A_k u_{ki} \tag{1}$$

yields

$$A(Bu_{ki}) = A_k(Bu_{ki}). \tag{2}$$

Thus Bu_{ki} is an eigenfunction of A belonging to the eigenvalue A_k; consequently

$$Bu_{ki} = \sum_{j=1}^{n} u_{kj}(u_{kj}, Bu_{ki}) = \sum_{j=1}^{n} u_{kj}B_{ji}. \tag{3}$$

In the absence of degeneracy (i.e., $n = 1$), u_k is a simultaneous eigenfunction of A and B with eigenvalues A_k and B_k respectively.

In discussing the degenerate case, the constant index k may be omitted. The equations

$$u_i' = \sum_{r=1}^{n} u_r T_{ri} \tag{4}$$

define a new orthonormal set if the transformation matrix T has the unitary property. This property arises from subjecting u_i' to the orthonormal condition

$$(u_i', u_j') = \delta_{ij}. \tag{5}$$

The calculation yields

$$\sum_{l=1}^{n} \overline{T_{li}}\, T_{lj} = \delta_{ij}. \tag{6}$$

Note that

$$T_{ji} = (u_j, u_i'). \tag{7}$$

Consequently

$$T^*_{ij} = \overline{T_{ji}} = (u_i', u_j) \tag{8}$$

and

$$u_j = \sum_{l=1}^{n} u_l' T^*_{lj}. \tag{9}$$

Interchanging the role of primed and unprimed functions in the calculation leading to Eq. (6) gives

$$\sum_{l=1}^{n} \overline{T^*_{li}}\, T^*_{lj} = \sum_{l=1}^{n} \overline{T_{jl}}\, T_{il} = \delta_{ij}. \tag{10}$$

We seek to determine the matrix T so that the new orthonormal functions u_i' defined by Eq. (4) are solutions of the eigenvalue problem

$$Bu_i' = B_i' u_i'. \tag{11}$$

Combining Eqs. (3) and (11), we get

$$\sum_j Bu_j T_{ji} = \sum_p \sum_j u_p B_{pj} T_{ji} = B_i' \sum_p u_p T_{pi},$$

or, omitting the second index i for the moment,

$$\sum_j B_{pj} T_j = B' T_p, \quad p = 1, 2, \ldots, n. \tag{12}$$

A necessary and sufficient condition for the existence of a nontrivial solution of Eq. (12) is the vanishing of the determinant of the coefficients:

$$\|B_{pq} - B'\delta_{pq}\| = 0. \tag{13}$$

The proof that B' in Eq. (12) is a real number is left as an exercise for the reader. The n real roots of Eq. (13) are the possible eigenvalues of the operator B in the function space u_i.

Returning to Eq. (12) with a particular root, we can solve it and interpret the normalized solution as a column of the matrix T. The actual complete construction of T and the eigenfunctions u_i' can be carried through by a repetitive process described in detail in Ref. 6 (a). The procedure is based upon the theorem that a succession of unitary transformations considered as a unit possesses the unitary property (another instructive exercise for the reader). The reader will also find it instructive to show that the roots of the determinantal equation are invariant under an arbitrary unitary transformation of the basis functions u_i.

In the primed orthonormal system, the matrix elements of B are

$$\begin{aligned}
B_{ij}' = B_i'\delta_{ij} &= (u_i', Bu_j') \\
&= \sum_{pq} (u_p, Bu_q)\, \overline{T_{pi}}\, T_{pj} \\
&= \sum_{pq} T_{ip}^* B_{pq} T_{qi},
\end{aligned} \tag{14}$$

or, as a matrix equation,

$$B' = T^*BT. \tag{15}$$

In problems of physical interest, whenever the eigenvalues of an operator or set of operators are degenerate it is possible to construct additional operators which commute with each other, and with the original set, such that the extended set of operators possesses only nondegenerate eigenvalues. This means that no two sets of eigenvalues of all the operators are identical. Such sets of operators are said to be *complete*. Let β denote a particular set of the eigenvalues; the normalized eigenfunction u_β is then uniquely determined except for an arbitrary phase factor of magnitude unity.

If β and β' are two different sets of eigenvalues corresponding to the respective eigenfunctions u_β and u_β', the matrix element of an arbitrary operator A may now be written in the Dirac notation:

$$(\beta|A|\beta') = (u_\beta, A u_\beta'). \tag{16}$$

Without proof, we state in Dirac notation the formal rule for expressing the matrix elements of a product operator AB in terms of the matrix elements of A and B:

$$(\beta|AB|\beta') = \sum_{\beta''} (\beta|A|\beta'')(\beta''|B|\beta') \tag{17}$$

1–5 An example: the harmonic oscillator in one dimension. A harmonic oscillator of mass M is supposed subject to a restoring force $-kq$. If angular momentum is measured in units of $\hbar$, the commutation rule is

$$pq - qp = -i. \tag{1}$$

If we measure energy in units of $\hbar(k/M)^{\frac{1}{2}}$, the Hamiltonian for the oscillator can be written

$$\mathfrak{K} = \tfrac{1}{2}(P^2 + Q^2), \tag{2}$$

where

$$\begin{aligned} P &= \hbar^{\frac{1}{2}}(Mk)^{-\frac{1}{4}}p, \\ Q &= \hbar^{-\frac{1}{2}}(Mk)^{\frac{1}{4}}q. \end{aligned} \tag{3}$$

The transformation (3) evidently preserves the form of Eq. (1):

$$PQ - QP = -i. \tag{4}$$

From Eq. (2) the eigenvalue equation is

$$(P^2 + Q^2)\psi = 2E\psi. \tag{5}$$

Now

$$P^2 + Q^2 = (P \pm iQ)(P \mp iQ) \pm 1 \tag{6}$$

and

$$\begin{aligned} (P \mp iQ)(P^2 + Q^2) &= (P \mp iQ)[(P \pm iQ)(P \mp iQ) \pm 1] \\ &= [(P \mp iQ)(P \pm iQ) \pm 1](P \mp iQ) \\ &= [P^2 + Q^2 \pm 2](P \mp iQ). \end{aligned} \tag{7}$$

Application of $P + iQ$ to both members of Eq. (5) transforms it into

$$[P^2 + Q^2 - 2]\{(P + iQ)\psi\} = 2E\{(P + iQ)\psi\},$$

or

$$(P^2 + Q^2)\{(P + iQ)\psi\} = 2(E + 1)\{(P + iQ)\psi\}. \tag{8}$$

Repetition of this process yields

$$(P^2 + Q^2)\{(P + iQ)^n\psi\} = 2(E + n)\{(P + iQ)^n\psi\}. \tag{9}$$

Thus, if ψ is an eigenfunction with the eigenvalue E, $(P + iQ)^n\psi$ is an eigenfunction with the eigenvalue $E + n$. The ladder of solutions defined in this manner extends upward indefinitely. The same procedure with the adjoint operator $P - iQ$ yields

$$(P^2 + Q^2)\{(P - iQ)^m\psi\} = 2(E - m)\{(P - iQ)^m\psi\}. \tag{10}$$

If m is sufficiently great $(E - m)$ becomes negative. But Eq. (10) then implies a negative diagonal matrix element for the operator $P^2 + Q^2$, in contradiction to the general relation of Eq. (10), Section 1–2. Consequently, $(P - iQ)^m\psi$ must vanish if $(E - m)$ is negative. If now ψ_0 is taken to be the eigenfunction corresponding to the smallest eigenvalue,

$$(P - iQ)\psi_0 = (1/i)\left(\frac{\partial}{\partial Q} + Q\right)\psi_0 = 0. \tag{11}$$

The solution of this equation is

$$\psi_0 = N_0 e^{-Q^2/2}$$

with the eigenvalue $E_0 = \frac{1}{2}$. For $E_n = n + \frac{1}{2}$, the corresponding eigenfunctions are

$$\psi_n = N_n\left(\frac{\partial}{\partial Q} - Q\right)^n e^{-Q^2/2}. \tag{12}$$

The nonvanishing matrix elements of $P \pm iQ$ can be derived from the relation

$$(n|P^2 + Q^2|n) = 2n + 1. \tag{13}$$

With the aid of Eq. (6) this becomes

$$(n|(P + iQ)(P - iQ)|n) = 2n. \tag{14}$$

The rule for the matrix element of a product operator [Eq. (17), Section 1–4] allows the reduction of Eq. (14) to

$$(n|P + iQ|n - 1)(n - 1|P - iQ|n) = 2n. \tag{15}$$

The infinite sum arising from the application of the product rule to Eq. (14) reduces to a single term because the displacement operators $P \pm iQ$ change the index n by one unit.

If we adopt the convention that the nonvanishing matrix elements of Q are real positive numbers, Eq. (15) yields

$$(n|Q - iP|n - 1) = (2n)^{\frac{1}{2}}, \tag{16}$$

since the Hermitian property of P and Q individually requires that

$$(n - 1|P - iQ|n) = \overline{(n|P|n - 1)} - i\overline{(n|Q|n - 1)}$$
$$= \overline{(n|P + iQ|n - 1)}. \tag{17}$$

Other matrix elements which follow from Eqs. (15) and (16) are

$$(n|Q + iP|n - 1) = 0,$$
$$(n - 1|Q + iP|n) = (2n)^{\frac{1}{2}},$$
$$(n - 1|Q - iP|n) = 0.$$
$$(n-1|Q|n) = \sqrt{\frac{n}{2}}$$
$$(n|P|n-1) = i\sqrt{\frac{n}{2}} \tag{18}$$

This example contains several features analogous to those of the angular momentum problem, as will be evident in Chapter 3. These are particularly the occurrence of displacement operators ($P \pm iQ$), the ladder of solutions, and the lower bound on the eigenvalues of an operator consisting of a sum of squares of Hermitian operators.

CHAPTER 2

COMMUTATION RULES

2–1 The fundamental commutator for angular momentum. The angular momentum of a group of particles is defined classically as the vector quantity

$$\mathbf{L}_{cl} = \sum_k \mathbf{r}_k \times \mathbf{p}_k, \tag{1}$$

in which $\mathbf{r}_k$ is the coordinate vector of the kth particle and $\mathbf{p}_k$ is the conjugate linear momentum. This quantity can be resolved into two terms, the angular momentum of the particles about the center of mass and the angular momentum of the center of mass about the fixed origin. In many applications the center of mass can be identified with a fixed origin (the nucleus in atomic problems, the center of a sphere containing all the nucleons in nuclear problems).

The quantum mechanical operator corresponding to $\mathbf{L}_{cl}$ is

$$\mathbf{L} = \sum_k \frac{\hbar}{i} \mathbf{r}_k \times \nabla_k \tag{2}$$

derived from $\mathbf{L}_{cl}$ by the prescription $\mathbf{p}_k \to (\hbar/i)\nabla_k$. Explicitly, in terms of rectangular components,

$$\begin{aligned}
\mathbf{L}_x &= \sum_k \frac{\hbar}{i}\left(y_k \frac{\partial}{\partial z_k} - z_k \frac{\partial}{\partial y_k}\right), \\
\mathbf{L}_y &= \sum_k \frac{\hbar}{i}\left(z_k \frac{\partial}{\partial x_k} - x_k \frac{\partial}{\partial z_k}\right), \\
\mathbf{L}_z &= \sum_k \frac{\hbar}{i}\left(x_k \frac{\partial}{\partial y_k} - y_k \frac{\partial}{\partial x_k}\right).
\end{aligned} \tag{3}$$

Note that L_y and L_z can be derived from L_x by a cyclic permutation of variables: $x \to y \to z \to x$. In applications it is convenient to use $\hbar$ as the unit of angular momentum. Then all angular momentum symbols I, L, S, etc., represent the physical operators divided by $\hbar$. In accordance with this convention, Eq. (3) implies the commutation relations

$$[L_y, L_z] = L_y L_z - L_z L_y = iL_x,$$
$$[L_z, L_x] = L_z L_x - L_x L_z = iL_y, \qquad (4)$$
$$[L_x, L_y] = L_x L_y - L_y L_x = iL_z,$$

which can be written briefly as

$$\mathbf{L} \times \mathbf{L} = i\mathbf{L}. \qquad (5)$$

This is the fundamental commutation rule for angular momentum. An interesting feature of the relation is that one cannot begin with it and proceed, through some kind of inverse process, to the explicit form for $\mathbf{L}$ from which the calculation started. The commutation relation (5) is of quite general form so far as angular momentum is concerned, and the same form applies to intrinsic angular momentum (or spin). Since spin has no classical counterpart, it cannot be expressed as a function of the canonically conjugate variables to which the prescription for forming operators may be applied. Instead, one assumes that the intrinsic angular momentum operator

$$\mathbf{S} = \sum_k \mathbf{S}_k \qquad (6)$$

obeys the same commutation rule that is found for orbital angular momentum,

$$\mathbf{S} \times \mathbf{S} = i\mathbf{S} \qquad (7)$$

and this assumption is then subjected to careful test (and verified) by experiment.

Just as the operator $\mathbf{S}$ has no classical counterpart, the spin coordinate space in which it operates is entirely independent of the ordinary space coordinates upon which $\mathbf{L}$ operates. Consequently the components of $\mathbf{L}$ commute with those of $\mathbf{S}$, and the total angular momentum $\mathbf{I} = \mathbf{L} + \mathbf{S}$ for a system of particles obeys the commutation rule

$$\mathbf{I} \times \mathbf{I} = i\mathbf{I}. \qquad (8)$$

The symbol $\mathbf{I}$, commonly applied to nuclear systems, is used instead of $\mathbf{J}$, which is reserved for atomic systems.

As with the material of Chapter 1, the brief mention of spin in this section is intended merely to refresh the reader's mind concerning the groundwork upon which succeeding chapters build. More detailed discussions are found in the standard references at the end of Chapter 5.

2-2 Angular momentum and rotations of axes. (a) *Rotation of axes.* Figure 1 indicates a rotation of coordinate axes about the

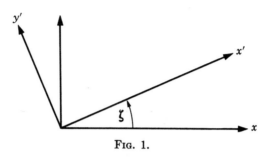

Fig. 1.

z-axis through an angle ζ. The primed and unprimed coordinates of a point are related through the equations

$$x' = x \cos \zeta + y \sin \zeta,$$
$$y' = -x \sin \zeta + y \cos \zeta, \qquad (1)$$
$$z' = z,$$

or

$$x' \pm iy' = (x \pm iy)e^{\mp i\zeta},$$
$$z' = z. \qquad (2)$$

We wish now to show that this rotation can be generated by a unitary operator constructed from angular momentum operators. Consider the operator

$$Z_\zeta = e^{i\zeta I_z}, \qquad (3)$$

by which is meant the usual exponential power series in the quantity $i\zeta I_z$.

Since I_z is Hermitian, the operator obtained by replacing i by $-i$ in Eq. (3) generates the adjoint of Z_ζ:

$$Z_\zeta^* = e^{-i\zeta I_z}. \qquad (4)$$

This statement can be verified by consideration of a typical term of the power series for Z_ζ in the integral relation [Eq. (8), Section 1-2] defining the adjoint operator. It is also clear from the power series definition of the exponential that the theorem $\exp(\alpha)\exp(\beta) =$

$\exp(\alpha + \beta)$ holds not only when α and β are numbers, but equally well when they are commuting operators. Consequently,

$$Z_\zeta^{-1} = Z_\zeta^* = e^{-i\zeta I_z}. \tag{5}$$

and Z_ζ possesses the unitary property.

Let us evaluate the effect of Z_ζ on the function $x + iy$ when I_z contains an addend of the form $\dfrac{1}{i}\left(x\dfrac{\partial}{\partial y} - y\dfrac{\partial}{\partial x}\right)$. From the commutator

$$[I_z, x \pm iy] = \left[\frac{1}{i}\left(x\frac{\partial}{\partial y} - y\frac{\partial}{\partial x}\right), x \pm iy\right] = \pm(x \pm iy), \tag{6}$$

we obtain

$$I_z^n(x \pm iy) = (x \pm iy)(\pm 1 + I_z)^n \tag{7}$$

and

$$e^{-i\zeta I_z}(x \pm iy) = (x \pm iy)e^{-i\zeta(\pm 1 + I_z)}. \tag{8}$$

Consequently,

$$e^{-i\zeta I_z}(x \pm iy)e^{i\zeta I_z} = (x \pm iy)e^{\mp i\zeta}. \tag{9}$$

It is evident that z is unaltered by the transformation, which thus is equivalent to a rotation of coordinates through the angle ζ about the z-axis as expressed by Eq. (2).

The effect of Z_ζ on an arbitrary function of particle coordinates $x_1, y_1, z_1, \ldots, x_n, y_n, z_n, \ldots$ is most easily discussed using spherical coordinates. Let

$$\begin{aligned}
x_n &= r_n \sin\theta_n \cos\phi_n, \\
y_n &= r_n \sin\theta_n \sin\phi_n, \\
z_n &= r_n \cos\theta_n,
\end{aligned} \tag{10}$$

and

$$\begin{aligned}
\beta &= \phi_1, \\
\beta_2 &= \phi_2 - \phi_1, \\
\beta_n &= \phi_n - \phi_1.
\end{aligned} \tag{11}$$

A change in the angle β while all other variables remain constant is equivalent to a rigid rotation of the particle system about the z-axis. Now

$$x\frac{\partial}{\partial y} - y\frac{\partial}{\partial x} = \frac{\partial}{\partial \phi} \tag{12}$$

and

$$L_z = \frac{1}{i} \sum_n \left(x_n \frac{\partial}{\partial y_n} - y_n \frac{\partial}{\partial x_n} \right)$$
$$= \frac{1}{i} \sum_n \frac{\partial}{\partial \phi_n} = \frac{1}{i} \frac{\partial}{\partial \beta}, \tag{13}$$

where the last equality of (13) follows from the relations

$$\frac{\partial}{\partial \phi_1} = \frac{\partial \beta}{\partial \phi_1} \frac{\partial}{\partial \beta} + \sum_{n>1} \frac{\partial \beta_n}{\partial \phi_1} \frac{\partial}{\partial \beta_n}$$
$$= \frac{\partial}{\partial \beta} - \sum_{n>1} \frac{\partial}{\partial \beta_n},$$
$$\frac{\partial}{\partial \phi_n} = \frac{\partial \beta_n}{\partial \phi_n} \frac{\partial}{\partial \beta_n} = \frac{\partial}{\partial \beta_n} \quad [n > 1].$$

If ψ is a function of the variables $r_1 .. r_n .. , \theta_1 .. \theta_n .. , \beta_2 .. \beta_n .. ; \beta$, it follows from Eq. (13) that

$$Z_\zeta^{-1} \psi Z_\zeta = e^{-\zeta(\partial/\partial\beta)} \psi e^{\zeta(\partial/\partial\beta)}. \tag{14}$$

But if we expand the right member, making use of the usual exponential power series, and evaluate the commutators indicated in the resulting expansion, we obtain simply a Taylor's expansion of ψ in the variable $-\zeta$ about the point β. Consequently,

$$\psi(r_1 .. r_n .. , \theta .. \theta_n .. , \beta_2 .. \beta_n .. ; \beta - \zeta)$$
$$= e^{-i\zeta I_z} \psi(r_1 .. r_n .. , \theta_1 .. \theta_n .. , \beta_2 .. \beta_n; \beta) e^{i\zeta I_z} \tag{15}$$

or, expressing ψ as a function of rectangular coordinates,

$$\psi(\ldots x_n', y_n', z_n' \ldots) = e^{-i\zeta I_z} \psi(\ldots x_n, y_n, z_n \ldots) e^{i\zeta I_z}. \tag{16}$$

The transformation equation (2) remains valid if the position vector x,y,z is replaced by an arbitrary vector function V_x, V_y, V_z. But Eq. (8) cannot be generalized in the same way for an arbitrary vector **V**. The necessary restrictions on **V** are made evident by examining the effect of an infinitesimal transformation. Consider left and right members of an equation like (8) for an arbitrary vector **V**. If ζ is treated as an arbitrarily small quantity, the left member reduces to

$$(1 - i\zeta I_z)(V_x \pm iV_y)(1 + i\zeta I_z)$$
$$= V_x \pm iV_y - i\zeta[I_z, V_x \pm iV_y]. \tag{17}$$

Similarly, the right member is

$$(V_x \pm iV_y)(1 \mp i\zeta); \qquad (18)$$

equality of these two expressions requires

$$[I_z, V_x \pm iV_y] = \pm(V_x \pm iV_y). \qquad (19a)$$

The same argument applied to V_z yields a requirement

$$[I_z, V_z] = 0. \qquad (19b)$$

Equation (19) and two others obtained from it by cyclic permutation of the x,y,z subscripts are satisfied by a large class of physically important vector functions, including the position, electric dipole moment, linear momentum, orbital angular momentum, intrinsic spin, and magnetic moment vector of a single particle.

The distinction between the algebraic description of a rotation of coordinate axes [Eq. (2)] and the transformation induced by the operator Z_ζ can now be clearly formulated: the operator does not rotate coordinate axes, but generates a rigid rotation of the physical system itself. Under the rigid rotation the coordinates of a particle are transformed in accordance with Eq. (2) as if the system were held fixed and the inverse rotation applied to the coordinate axes. The components V_x, V_y, V_z of a vector quantity which participates in the rigid rotation of the physical system transform into V_x', V_y', V_z' exactly as required by the transformation law relating the components of a vector in different orthogonal coordinate systems. On the other hand, the components of a constant vector (as, for example, a constant external magnetic field) are unchanged by the operator Z_ζ.

Equation (19) is a basic relation in the derivation of selection rules, transition probabilities, and energy displacements for systems possessing a center of symmetry.

It should perhaps be noted here that an arbitrary infinitesimal rotation may be compounded from three infinitesimal rotations through angles ξ, η, ζ about the three rectilinear axes. The order in which the rotations are applied is immaterial.

(b) *Eigenvalues of orbital angular momentum.* A brief excursion outside the topic of this section is suggested by the relation (13). The eigenvalue equation

$$L_z \psi(\ldots x_n, y_n, z_n \ldots) = \mu \psi(\ldots x_n, y_n, z_n \ldots) \qquad (20)$$

may be solved by writing ψ as a function of the spherical coordinates defined by Eqs. (10) and (11), i.e.,

$$\frac{1}{i} \frac{\partial}{\partial \beta} \psi(r_1 \ldots r_n \ldots, \theta_1 \ldots \theta_n \ldots, \beta_2 \ldots \beta_n \ldots ; \beta)$$

$$= \mu \psi(r_1 \ldots r_n \ldots, \theta_1 \ldots \theta_n \ldots, \beta_2 \ldots, \beta_n \ldots ; \beta). \quad (21)$$

From Eq. (21) it follows that ψ depends upon β only through a factor $\exp(i\mu\beta)$. The condition that the wave function be a single-valued function of the particle coordinates restricts μ to integral values $0, \pm 1, \ldots, \pm m, \ldots$. These are the eigenvalues of a single component of the orbital angular momentum. For one particle the angle functions determined in this manner are the surface spherical harmonics of integral order. The fact that these functions form a complete set justifies the restriction to single-valued point functions.

2–3 Invariance of the Hamiltonian and of I^2 under rotations. Consider the Hamiltonian function

$$\mathcal{K} = \sum_n \frac{1}{2m_n} (p_{xn}^2 + p_{yn}^2 + p_{zn}^2) + V(\ldots x_n, y_n, z_n \ldots) \quad (1)$$

associated with an isolated system of particles. The operator corresponding to Eq. (1) is

$$\mathcal{K} = -\sum \frac{\hbar^2}{2m_n} \Delta_n + V(\ldots x_n, y_n, z_n \ldots). \quad (2)$$

The algebraic relations connecting the coordinates of a point $(x, y, z$ and $x', y', z')$ in two reference frames related by a rigid rotation of coordinate axes yield an identity

$$\frac{\partial^2}{\partial x'^2} + \frac{\partial^2}{\partial y'^2} + \frac{\partial^2}{\partial z'^2} = \frac{\partial^2}{\partial x^2} + \frac{\partial^2}{\partial y^2} + \frac{\partial^2}{\partial z^2}. \quad (3)$$

This relation is perhaps more familiarly formulated as the invariance of the scalar product $\nabla' \cdot \nabla' = \nabla \cdot \nabla$. An equivalent statement in terms of the angular momentum operators is

$$e^{-i\mathcal{S}I_z} \Delta e^{i\mathcal{S}I_z} = \Delta, \quad (4)$$

or, more generally and simply,

$$[\mathbf{L}, \Delta] = 0. \quad (5)$$

To verify Eq. (5) we evaluate the effect of the operator $i(L_z\Delta - \Delta L_z)$ on an arbitrary function ψ:

$$i(L_z\Delta - \Delta L_z)\psi$$
$$= \left[\left(x\frac{\partial}{\partial y} - y\frac{\partial}{\partial x}\right)\left(\frac{\partial^2}{\partial x^2} + \frac{\partial^2}{\partial y^2}\right) - \left(\frac{\partial^2}{\partial x^2} + \frac{\partial^2}{\partial y^2}\right)\left(x\frac{\partial}{\partial y} - y\frac{\partial}{\partial x}\right)\right]\psi$$
$$= -2\left[\frac{\partial}{\partial x}\frac{\partial}{\partial y} - \frac{\partial}{\partial y}\frac{\partial}{\partial x}\right]\psi = 0.$$

What is the effect of rotations on the potential function V? The general answer for an isolated system is that V must be invariant under an arbitrary rigid rotation of the system. Failure of this condition would single out certain orientations as having peculiar properties, i.e., would require an influence of the external world on the system, contrary to the isolation postulate. Thus all physically admissible potential energy functions for isolated systems obey the commutation relation

$$[\mathbf{I},V] = 0. \tag{6}$$

Equations (5) and (6) together yield

$$[\mathbf{I},\mathfrak{K}] = 0. \tag{7}$$

A special case of great importance is that in which V is a function of space coordinates only. Then

$$[\mathbf{I},V] = [\mathbf{L},V]$$
$$= -i\sum_n \mathbf{r}_n \times (\boldsymbol{\nabla}_n V) \tag{8}$$
$$= i\ (\text{torque}),$$

where the interpretation of the sum in Eq. (8) as the net torque on the system is based upon recognition of $-\boldsymbol{\nabla}_n V$ as the classical force on the nth particle. If the net torque vanishes, V is invariant under a rigid rotation, and conversely.

An example to which the foregoing remarks apply is the electrostatic potential function of a system of charged particles. In this case V depends only upon the invariant quantities

$$r_k = \sqrt{x_k^2 + y_k^2 + z_k^2},$$
$$r_{kl} = \sqrt{(x_k - x_l)^2 + (y_k - y_l)^2 + (z_k - z_l)^2}, \tag{9}$$

the origin being at the nucleus.   More generally, V may depend on interparticle distances, the lengths of vectors physically associated with the particles, the scalar products of such vectors, or other quantities having values entirely independent of the orientation of the systems.

The invariance of I^2 under rotations is quickly established by evaluating the commutator of $I^2 = I_x^2 + I_y^2 + I_z^2$ with a component of I, say I_z:

$$\begin{aligned}
[I^2, I_z] &= (I_x^2 + I_y^2)I_z - I_z(I_x^2 + I_y^2) \\
&= I_x(I_xI_z - I_zI_x) - (I_zI_x - I_xI_z)I_x \\
&\quad + I_y(I_yI_z - I_zI_y) - (I_zI_y - I_yI_z)I_y, \\
[I^2, I_z] &= -i(I_xI_y + I_yI_x) + i(I_yI_x + I_xI_y) \\
&= 0.
\end{aligned} \tag{10}$$

The derivation of

$$[I^2, \mathcal{3C}] = 0 \tag{11}$$

from Eq. (7) is left to the reader as an elementary exercise.

CHAPTER 3

THE EIGENVALUES AND MATRIX ELEMENTS OF ANGULAR MOMENTUM

3-1 The eigenvalues of I_z and $\mathbf{I}^2$. Let us denote by $\psi(\ldots |\alpha\lambda\mu)$ a simultaneous eigenfunction of $\mathcal{K}$, $\mathbf{I}^2$, and I_z, where α, λ, and μ are the respective eigenvalues of these operators:

$$\mathcal{K}\psi(\ldots |\alpha\lambda\mu) = \alpha\psi(\ldots |\alpha\lambda\mu), \tag{1}$$
$$\mathbf{I}^2\psi(\ldots |\alpha\lambda\mu) = \lambda\psi(\ldots |\alpha\lambda\mu), \tag{2}$$
$$I_z\psi(\ldots |\alpha\lambda\mu) = \mu\psi(\ldots |\alpha\lambda\mu). \tag{3}$$

In discussing λ and μ, we shall find it convenient to employ the operators

$$M_{\backprime} = I_x - iI_y,$$
$$N = I_x + iI_y,$$

instead of the individual components I_x and I_y. The fundamental commutation rules for M and N follow from Eq. (8), Section 2-1:

$$I_zN - NI_z = N,$$
$$I_zM - MI_z = -M, \tag{4}$$
$$NM - MN = 2I_z.$$

Consider first Eq. (3) and apply N from the left:

$$NI_z\psi(\ldots |\alpha\lambda\mu) = \mu N\psi(\ldots |\alpha\lambda\mu). \tag{5}$$

The first line of Eq. (4) asserts that $NI_z = (I_z - 1)N$, which may be substituted directly into Eq. (5) to obtain

$$(I_z - 1)N\psi(\ldots |\alpha\lambda\mu) = \mu N\psi(\ldots |\alpha\lambda\mu),$$

which is equivalent to

$$I_z[N\psi(\ldots |\alpha\lambda\mu)] = (\mu + 1)N\psi(\ldots |\alpha\lambda\mu). \tag{6}$$

Therefore $N\psi(\ldots |\alpha\lambda\mu)$ is an eigenfunction of I_z with an eigenvalue $\mu + 1$. Precisely similar treatment of Eq. (3) with the operator M reveals that $M\psi(\ldots |\alpha\lambda\mu)$ is an eigenfunction of I_z with an eigenvalue $\mu - 1$. Because both M and N commute with $\mathcal{K}$ and $\mathbf{I}^2$, the functions $N\psi(\ldots|\alpha\lambda\mu)$ and $M\psi(\ldots|\alpha\lambda\mu)$ are simultaneous eigenfunctions of $\mathfrak{IC}$ and I^2 with the same eigenvalues α and λ which belong to the original $\psi(\ldots|\alpha\lambda\mu)$.

The displacement operators M and N can be used to generate new eigenfunctions of I_z, starting from $\psi(\ldots|\alpha\lambda\mu)$, with eigenvalues displaced by integral amounts in either direction from the initial eigenvalue μ. For example, $N\psi(\ldots|\alpha\lambda\mu)$ and $N^2\psi(\ldots|\alpha\lambda\mu)$ correspond to eigenvalues $\mu + 1$ and $\mu + 2$, whereas $M^3\psi(\ldots|\alpha\lambda\mu)$ has an eigenvalue $\mu - 3$. Thus we can construct for the eigenvalues of Eq. (3) a "ladder" of solutions analogous to that obtained for the one-dimensional oscillator in Section 1–5, with unit spacing between adjacent rungs:

$$
\begin{array}{ll}
\vdots & \\
N^3\psi: \underline{\hspace{2cm}} & \mu + 3 \\
N^2\psi: \underline{\hspace{2cm}} & \mu + 2 \\
N\psi: \underline{\hspace{2cm}} & \mu + 1 \\
\psi: \underline{\hspace{2cm}} & \mu \\
M\psi: \underline{\hspace{2cm}} & \mu - 1 \\
M^2\psi: \underline{\hspace{2cm}} & \mu - 2 \\
\vdots &
\end{array}
$$

As the ladder is shown, it appears to extend indefinitely in either direction from μ, but such is not the case. The number of rungs is finite and is related in a precise way to the eigenvalue λ of I^2. To demonstrate this, we form the product MN:

$$MN = I_x^2 + I_y^2 + i(I_xI_y - I_yI_x). \qquad (7)$$

The fundamental commutation rule [Eq. (8), Section 2–1] may be applied to Eq. (7) and, since $I_x^2 + I_y^2 = I^2 - I_z^2$, it follows that

$$MN = I^2 - I_z^2 - I_z. \qquad (8)$$

Equation (8) reveals that an eigenfunction of I^2 and I_z is also an eigenfunction of MN. Let $\psi(\ldots|\alpha\lambda\mu)$ be any such eigenfunction. Then

$$MN\psi(\ldots|\alpha\lambda\mu) = (\lambda - \mu^2 - \mu)\psi(\ldots|\alpha\lambda\mu). \qquad (9)$$

Our argument to show that the ladder terminates in both directions is based upon the corollary, proved below, that the eigenvalues of MN given by Eq. (9) cannot be negative.

The eigenvalue of MN for a state ψ can be expressed as the matrix element

$$\int \cdots \int \bar{\psi}(I_x - iI_y)(I_x + iI_y)\psi \, d\tau. \qquad (10)$$

Since the individual components of $\mathbf{I}$ are Hermitian, that is, since

$$\int \cdots \int \bar{\psi}I_x\phi \, d\tau = \int \cdots \int \overline{(I_x\psi)}\phi \, d\tau,$$

it follows that $I_x - iI_y$ is not Hermitian. In fact, direct expansion using the Hermitian properties of I_x and I_y verifies that

$$\int \cdots \int \bar{\psi}(I_x - iI_y)(I_x + iI_y)\psi \, d\tau$$

$$= \int \cdots \int \overline{[(I_x + iI_y)\bar{\psi}]}(I_x + iI_y)\psi \, d\tau$$

$$= \int \cdots \int |(I_x + iI_y)\psi|^2 \, d\tau. \qquad (11)$$

The integral (11) is necessarily non-negative, as therefore are the eigenvalues of MN. But these eigenvalues are found from Eq. (9), and it follows that

$$\lambda - \mu^2 - \mu \geq 0. \qquad (12)$$

Now $\lambda = (\psi, I_x^2\psi) + (\psi, I_y^2\psi) + (\psi, I_z^2\psi)$ is a sum of non-negative numbers, and the expression

$$\lambda \geq \mu^2 + \mu \qquad (13)$$

therefore places an upper limit on the absolute value of μ.

The fact that the ladder terminates at its upper end permits us to evaluate λ in terms of $\mu_{\max}$. Since no eigenfunction exists with eigenvalue greater than $\mu_{\max}$, $N\psi(\ldots |\alpha\lambda\mu_{\max})$ cannot generate an eigenfunction with $\mu_{\max} + 1$ as its eigenvalue, and an equation like (6) can be valid for $N\psi(\ldots |\alpha\lambda\mu_{\max})$ only if $N\psi(\ldots |\alpha\lambda\mu_{\max})$ vanishes. Then

$$MN\psi(\ldots |\alpha\lambda\mu_{\max}) = 0 = (\lambda - \mu_{\max}^2 - \mu_{\max})\psi(\ldots |\alpha\lambda\mu_{\max}) \quad (14)$$

and

$$\lambda = \mu_{\max}^2 + \mu_{\max}, \qquad (15)$$

where $\mu_{\max}$ is positive.

Similarly, beginning with NM instead of MN as in Eq. (7), we find that a lower limit to the ladder of eigenvalues exists, and that

$$\lambda = \mu_{min}^2 - \mu_{min}. \tag{16}$$

Therefore $\mu_{min} = -\mu_{max}$, and the eigenvalues form a symmetrical array about the origin.

It is usual to let $\mu = m$ and $\mu_{max} = I$. Thus the maximum value of m is I, the minimum being $-I$. Then Eq. (15) reveals that

$$\lambda = I(I + 1). \tag{17}$$

Instead of $I(I + 1)$, I is customarily used to label the eigenfunction:

$$\psi(\ldots |\alpha\lambda\mu) \equiv \psi(\ldots |\alpha Im). \tag{18}$$

The effect of N or M on $\psi(\ldots |\alpha Im)$ is then to generate an eigenfunction of I_z with eigenvalue $m + 1$ or $m - 1$ respectively.

The symmetry of the eigenvalues m about zero combines with the fact that adjacent eigenvalues differ by unity to require that the total number of distinct values of m is $2I + 1$. Thus $2I + 1$ is an integer, and I takes on values which are either integral or integral plus one-half. We have shown (Section 2–2) that orbital angular momenta necessarily have integral m-values. Therefore integral plus one-half values of I must describe spin states or states with compounded spin and orbital angular momenta.

Consider now a compound system with noninteracting components a and b described by

$$\psi_a(\ldots |\alpha I_a m_a) \quad \text{and} \quad \psi_b(\ldots |\beta I_b m_b).$$

The possible states of the compound system are given by the product functions

$$\psi_a(\ldots |\alpha I_a m_a)\psi_b(\ldots |\beta I_b m_b).$$

These functions are already eigenfunctions of $I_z = I_{za} + I_{zb}$ with eigenvalues $m = m_a + m_b$.

We wish to determine the possible eigenvalues of $\mathbf{I}^2 = (\mathbf{I}_a + \mathbf{I}_b)^2$ in the space of the product functions. This is done most simply by listing all pairs of values of m_a, m_b in the order of decreasing $m = m_a + m_b$ as shown in Table 1.

TABLE 1

$$I_a \geqq I_b$$

m_a	m_b	$m = m_a + m_b$	I
I_a	I_b	$I_a + I_b$	$I_a + I_b$
I_a	$I_b - 1$	$I_a + I_b - 1$	$I_a + I_b - 1$
$I_a - 1$	I_b		
I_a	$I_b - 2$	$I_a + I_b - 2$	$I_a + I_b - 2$
$I_a - 1$	$I_b - 1$		
$I_a - 2$	I_b		
$\vdots$	$\vdots$	$\vdots$	$\vdots$
I_a	$-I_b$	$I_a - I_b$	$I_a - I_b$
$I_a - 1$	$-I_b + 1$		
$I_a - 2$	$-I_b + 2$		
$\vdots$	$\vdots$		
$I_a - 2I_b$	$+I_b$		

The function space defined by $m = I_a + I_b$ contains only one product function. If $N = N_a + N_b$ is applied to this function it vanishes. Hence it is an eigenfunction of I^2 with the eigenvalue $I = I_a + I_b$. In the function space defined by $m = I_a + I_b - 1$ there are two linearly independent functions. A particular linear combination of these functions belongs to $I = I_a + I_b$; another linear combination orthogonal to the first belongs to the eigenvalue $I = I_a + I_b - 1$. In each successive function space there is one linear combination left over after the substates with the given m belonging to all previously occurring values of I have been accounted for. This remainder function space is thus necessarily an eigenfunction of I^2, with I equal to the m-value characterizing the function space. The process continues down to $I = I_a - I_b$; beyond that no possibilities occur for generating smaller values of I. Thus the eigenvalues of I range from $I_a + I_b$ down to $I_a - I_b$ in integral steps. Each value occurs just once.

3-2 The matrix elements of M, N, and I_z. The equivalent form for MN [Eq. (8), Section 3–1] permits determination of the effect of MN applied to $\psi(\ldots |\alpha I m)$:

$$MN\psi(\ldots |\alpha I m) = \{I(I + 1) - m(m + 1)\}\psi(\ldots |\alpha I m). \tag{1}$$

Upon forming the scalar product of $\psi(\ldots |\alpha I m)$ with $MN\psi$ we obtain the diagonal matrix element of MN:

$$(\alpha I m|MN|\alpha I m) = (I - m)(I + m + 1). \tag{2}$$

All nondiagonal elements vanish, as does the diagonal element for $m = I$, in accordance with our previous finding that N applied to a state with $m = I$ must give zero.

The matrix element of MN is also expressible by the matrix product rule as

$$(\alpha I m|MN|\alpha I m) = \sum_{m'} (\alpha I m|M|\alpha I m')(\alpha I m'|N|\alpha I m). \tag{3}$$

In the general application of the product rule, a sum over all possible intermediate states α' and I', as well as m', would be necessary. However, M and N commute with $\mathcal{3C}$ and $\mathbf{I}^2$. Consequently, the matrix elements of M and N are diagonal in α and I, and the sums over the intermediate states reduce to single terms.

Moreover, the sum over m' also reduces to a single term. As pointed out in the preceding section, the effect of M is to displace m downward by a unit, whereas N displaces m upward by a unit. Therefore the first matrix element in the right member of Eq. (3) is nonvanishing only if $m' = m + 1$, and the second is nonvanishing only if $m = m' - 1$. Equation (3) becomes simply

$$(\alpha I m|MN|\alpha I m) = (\alpha I m|M|\alpha I m + 1)(\alpha I m + 1|N|\alpha I m). \tag{4}$$

But the first element of the right member is the complex conjugate of the second. To see this, recall the Hermitian properties of I_x and I_y individually ($I_x \pm iI_y$ is *not* Hermitian):

$$
\begin{aligned}
\overline{(\alpha I m + 1|N|\alpha I m)} &= \overline{(\alpha I m + 1|I_x|\alpha I m)} + \overline{[i(\alpha I m + 1|I_y|\alpha I m)]} \\
&= (\alpha I m|I_x|\alpha I m + 1) + (-i)(\alpha I m|I_y|\alpha I m + 1) \\
&= (\alpha I m|M|\alpha I m + 1).
\end{aligned}
\tag{5}
$$

Then Eq. (4) becomes

$$(\alpha Im|MN|\alpha Im) = |(\alpha Im + 1|N|\alpha Im)|^2, \qquad (6)$$

which combines with Eq. (2) to yield

$$|(\alpha Im + 1|N|\alpha Im)|^2 = (I - m)(I + m + 1) \qquad (7)$$

or

$$(\alpha Im + 1|N|\alpha Im) = [(I - m)(I + m + 1)]^{\frac{1}{2}}e^{i\gamma_m}, \qquad (8)$$

where γ_m is a phase factor. It is conventional to adopt phases such that γ_m is zero. Then

$$(\alpha Im + 1|N|\alpha Im) = [(I - m)(I + m + 1)]^{\frac{1}{2}}. \qquad (9)$$

The matrix elements of I_x and I_y are obtained by recalling that, since M steps the eigenvalue m downward,

$$(\alpha Im + 1|M|\alpha Im) = 0. \qquad (10)$$

The sum and the difference of Eqs. (9) and (10) provide the required matrix elements of I_x and I_y, which are placed in the box below along with the only nonvanishing matrix elements of M and N as obtained from Eqs. (5) and (9). [See Table 2.]

TABLE 2

Nonvanishing Matrix Elements of I_x, I_y, M, and N	
$(\alpha Im + 1|I_x|\alpha Im) = \dfrac{1}{2}[(I - m)(I + m + 1)]^{\frac{1}{2}}$	
$(\alpha Im|I_x|\alpha Im + 1) = \dfrac{1}{2}[(I - m)(I + m + 1)]^{\frac{1}{2}}$	
$(\alpha Im + 1|I_y|\alpha Im) = \dfrac{-i}{2}[(I - m)(I + m + 1)]^{\frac{1}{2}}$	(11)
$(\alpha Im|I_y|\alpha Im + 1) = \dfrac{i}{2}[(I - m)(I + m + 1)]^{\frac{1}{2}}$	
$(\alpha Im + 1|N|\alpha Im) = [(I - m)(I + m + 1)]^{\frac{1}{2}}$ $(\alpha Im|M|\alpha Im + 1) = [(I - m)(I + m + 1)]^{\frac{1}{2}}$	(12)

The Hermitian character of I_x and I_y has been used to obtain the second and fourth elements listed above.

Finally, by forming the products $(\psi, I_z \psi)$, we obtain the only non-vanishing matrix elements of I_z as

$$(\alpha I m | I_z | \alpha I m) = m \tag{13}$$

3-3 The Pauli spin operators. As a particular example of the angular momentum matrices just discussed, we treat a system characterized by $I = \frac{1}{2}$. These matrices describe a state with intrinsic angular momentum only, since orbital angular momenta have integral I values. The nonvanishing matrix elements are quickly written from Eqs. (11), (12), and (13) of Section 3-2. (Only the magnetic quantum number m is carried explicitly, since all elements are diagonal in all other quantum numbers.)

$$
\begin{aligned}
[\tfrac{1}{2} | I_x | -\tfrac{1}{2}] &= \tfrac{1}{2}, \\
[-\tfrac{1}{2} | I_x | \tfrac{1}{2}] &= \tfrac{1}{2}, \\
[\tfrac{1}{2} | I_y | -\tfrac{1}{2}] &= -\frac{i}{2}, \\
[-\tfrac{1}{2} | I_y | \tfrac{1}{2}] &= \frac{i}{2}, \\
[\tfrac{1}{2} | I_z | \tfrac{1}{2}] &= \tfrac{1}{2}, \\
[-\tfrac{1}{2} | I_z | -\tfrac{1}{2}] &= -\tfrac{1}{2}.
\end{aligned}
\tag{1}
$$

As a matter of convenience the Pauli spin operator

$$\boldsymbol{\sigma} = 2\mathbf{I} \tag{2}$$

is introduced, whereupon the matrices for σ_x, σ_y, and σ_z can be written from Eq. (1) as

$$
\sigma_x = \begin{pmatrix} 0 & 1 \\ 1 & 0 \end{pmatrix},
$$
$$
\sigma_y = \begin{pmatrix} 0 & -i \\ i & 0 \end{pmatrix},
\tag{3}
$$
$$
\sigma_z = \begin{pmatrix} 1 & 0 \\ 0 & -1 \end{pmatrix};
$$

$m = \frac{1}{2}$ denotes the first row or the first column, and $m = -\frac{1}{2}$ denotes the second row or the second column.

The effects of the operator σ can be described using a variable which takes on just two values, corresponding to the two possible values of m when $I = \frac{1}{2}$. Then any function of the spin coordinates can be written

$$f(m) = f(\tfrac{1}{2})\, \delta_{\frac{1}{2}}(m) + f(-\tfrac{1}{2})\, \delta_{-\frac{1}{2}}(m), \qquad (4)$$

where $\delta_{m'}(m)$ is zero if $m' \neq m$ and has the value unity if $m' = m$. Thus $\delta_{m'}(m)$ can be considered as describing a state with $m = m'$, since it vanishes for other values of m. Matrix elements are formed according to the general rule

$$(\gamma'|F|\gamma'') = \int \cdots \int \overline{\psi(\gamma')} F \psi(\gamma)\, d\tau, \qquad (5)$$

and the analogous form yields matrix elements of σ as

$$(m'|\sigma|m'') = \sum_{m} \delta_{m'}(m)\sigma\, \delta_{m''}(m), \qquad (6)$$

where the summation allows the spin coordinates m to range over all possible values in its two-point space just as the coordinates τ of Eq. (5) range over all values in the integration.

In order to obtain the matrix elements of Eq. (3) from Eq. (6), it is necessary to define the operation of the components of σ on the δ's. For example, since I_z operates on $\psi(\ldots|\alpha I m)$ to produce a state with the same m,

$$\sigma_z\, \delta_{m'}(m) = 2m'\, \delta_{m'}(m). \qquad (7)$$

The operations of σ in the spin coordinate space are summarized below; the reader can quickly verify that these are consistent with Eqs. (6) and (3).

$$
\begin{aligned}
\sigma_z\, \delta_{\frac{1}{2}}(m) &= \delta_{\frac{1}{2}}(m), \\
\sigma_z\, \delta_{-\frac{1}{2}}(m) &= -\delta_{-\frac{1}{2}}(m), \\
\sigma_x\, \delta_{\frac{1}{2}}(m) &= \delta_{-\frac{1}{2}}(m), \\
\sigma_x\, \delta_{-\frac{1}{2}}(m) &= \delta_{\frac{1}{2}}(m), \\
\sigma_y\, \delta_{\frac{1}{2}}(m) &= i\, \delta_{-\frac{1}{2}}(m), \\
\sigma_y\, \delta_{-\frac{1}{2}}(m) &= -i\, \delta_{\frac{1}{2}}(m).
\end{aligned}
\qquad (8)
$$

The commutation relations satisfied by the components of $\boldsymbol{\sigma}$ follow readily from Eq. (8). Thus, for example,

$$\begin{aligned}
\sigma_x[\sigma_y\,\delta_{\frac{1}{2}}(m)] &= (\sigma_x\sigma_y)\,\delta_{\frac{1}{2}}(m) \\
&= i\sigma_x\,\delta_{-\frac{1}{2}}(m) \\
&= i\,\delta_{\frac{1}{2}}(m),
\end{aligned}$$

$$\begin{aligned}
\sigma_x[\sigma_y\,\delta_{-\frac{1}{2}}(m)] &= (\sigma_x\sigma_y)\,\delta_{-\frac{1}{2}}(m) \\
&= -i\sigma_x\,\delta_{\frac{1}{2}}(m) \\
&= -i\,\delta_{-\frac{1}{2}}(m),
\end{aligned}$$

or

$$\begin{aligned}
\sigma_x\sigma_y &= i\sigma_z, \\
\sigma_y\sigma_z &= i\sigma_x, \\
\sigma_z\sigma_x &= i\sigma_y.
\end{aligned} \tag{9}$$

Also, $\sigma_x\sigma_y = -\sigma_y\sigma_x$, etc., so that these relations are a particularly simple case of the general commutation relations. It is easily verified that

$$\sigma_x^2 = \sigma_y^2 = \sigma_z^2 = 1, \quad \sigma^2 = 3. \tag{10}$$

With the aid of Eqs. (9) and (10) any power series in the components of $\boldsymbol{\sigma}$ can be reduced to a linear function of these components. The rotation operator Z_ζ is an interesting example:

$$\begin{aligned}
Z_\zeta &\equiv e^{i\frac{1}{2}\zeta\sigma_z} \\
&= \cos\tfrac{1}{2}\zeta + i\sigma_z\sin\tfrac{1}{2}\zeta.
\end{aligned}$$

In the same way,

$$\begin{aligned}
Z_\xi &\equiv e^{i\frac{1}{2}\xi\sigma_x} \\
&= \cos\tfrac{1}{2}\xi + i\sigma_x\sin\tfrac{1}{2}\xi, \\
Z_\eta &\equiv e^{i\frac{1}{2}\eta\sigma_y} \\
&= \cos\tfrac{1}{2}\eta + i\sigma_y\sin\tfrac{1}{2}\eta.
\end{aligned}$$

These operators are double-valued functions of the angles. A rotation through 360 degrees restores the initial configuration of axes, but reverses the sign of the corresponding operator.

CHAPTER 4

MATRIX ELEMENTS OF SCALAR, VECTOR AND TENSOR OPERATORS

4–1 Vector-angular momentum commutation relations and matrix elements. The matrix elements of scalar and vector operators are often required in a representation in which one or more of the sets $(\mathbf{I}^2, I_z)$, $(\mathbf{L}^2, L_z)$, and $(\mathbf{S}^2, S_z)$ are diagonal. A familiar example is the spin orbit interaction, where the problem is to obtain matrix elements of the scalar operator $\mathbf{L} \cdot \mathbf{S}$ compounded from the independent vector operators $\mathbf{L}$ and $\mathbf{S}$ for a system in which $(\mathbf{L} + \mathbf{S})^2$ and $L_z + S_z$ are constants of the motion.

Both $\mathbf{L}$ and $\mathbf{S}$ belong to the class of vector operators to which the commutation relations of Eq. (19), Section 2–2, apply. This class is generated by linear combinations and vector products of the coordinate, momentum, and spin operators of the individual particles.

(a) *Commutation relations.* To secure the generality required by the diverse physical applications, we rewrite Eq. (19), Section 2–2, as a set of commutation relations among the components of a vector operator $\mathbf{U}$ and an angular momentum operator $\mathbf{K}$:

$$
\begin{aligned}
&(1) \ \ K_x U_x - U_x K_x = 0, \\
&(2) \ \ K_x U_y - U_y K_x = iU_z, \\
&(3) \ \ K_x U_z - U_z K_x = -iU_y. \\[4pt]
&(4) \ \ K_y U_y - U_y K_y = 0, \\
&(5) \ \ K_y U_z - U_z K_y = iU_x, \\
&(6) \ \ K_y U_x - U_x K_y = -iU_z. \\[4pt]
&(7) \ \ K_z U_z - U_z K_z = 0, \\
&(8) \ \ K_z U_x - U_x K_z = iU_y, \\
&(9) \ \ K_z U_y - U_y K_z = -iU_x.
\end{aligned}
\tag{1}
$$

A particular line of (1) will be referred to by the number preceding it.

The matrix elements of $\mathbf{U}$ are required in a representation in which $\mathbf{K}^2$ and K_z are diagonal; $\mathbf{K}$ is itself a possible $\mathbf{U}$ with matrix elements given by Eqs. (11), (12), and (13) of Section 3–2 on substituting the quantum number K for I. Possible operators **K** and **U** for which the commutation rules (1) hold are listed in Table 3 as Theorem I:

THEOREM I.   TABLE 3

If **K** is	then permissible **U**'s include
I	L,S,r_j,p_j,L_j,S_j
L	L_j,r_j,p_j
S	S_j
$I_j = L_j + S_j$	L_j,S_j,r_j,p_j

The theorem is quickly verified for any particular **K** and **U** using the fundamental rules and procedures of Chapter 2.

COROLLARY I–1. *If* **U** *and* **U**′ *both satisfy conditions* (1) *for the same* **K**, *then* **U**″ = **U**′ × **U** *also satisfies conditions* (1) *for that* **K**.

An example of the corollary is contained in Theorem I, since, for **K** = **L**, both r_j and p_j are acceptable **U**'s, and $L_j = r_j \times p_j$ is also an acceptable **U** of the theorem.

The fact that the components of **U**, **U**′, and **U**″ all transform in the same manner under rigid rotations of the physical system constitutes an adequate proof of the corollary. However, an explicit calculation is useful to test the consistency of the formalism.

We illustrate the calculation by establishing line (1) of Eq. (1) for **U**″ = **U**′ × **U**. By hypothesis,

$$K_x U_y - U_y K_x = iU_z,$$
$$K_x U_z - U_z K_x = -iU_y. \tag{2}$$

Applying U'_z from the left to the first equation, and U'_y from the left to the second equation, we have

$$U'_z K_x U_y - U'_z U_y K_x = iU'_z U_z,$$
$$U'_y K_x U_z - U'_y U_z K_x = -iU'_y U_y, \tag{3}$$

or

$$(K_x U'_z + iU'_y)U_y - U'_z U_y K_x = iU'_z U_z,$$
$$(K_x U'_y - iU'_z)U_z - U'_y U_z K_x = -iU'_y U_y. \tag{4}$$

The difference between the first and second lines of (4) is

$$K_x(U'_z U_y - U'_y U_z) - (U'_z U_y - U'_y U_z)K_x = 0, \tag{5}$$

which is line (1) of Eq. (1) for $\mathbf{U}'' = \mathbf{U}' \times \mathbf{U}$.

COROLLARY I–2.   *If* $\mathbf{U}$ *and* $\mathbf{K}$ *satisfy Eqs.* (1), *then* $\mathbf{K} \cdot \mathbf{U}$ ($=\mathbf{U} \cdot \mathbf{K}$) *commutes with* $\mathbf{K}$ (*and consequently with* $\mathbf{K}^2$).

The fact that $\mathbf{K} \cdot \mathbf{U}$ is invariant under rigid rotations of the physical system constitutes an adequate proof.   However, once more we test the consistency of the formalism by an explicit calculation.

Consider the commutator

$$K_x(K_x U_x + K_y U_y + K_z U_z) - (K_x U_x + K_y U_y + K_z U_z)K_x. \tag{6}$$

The terms containing only subscripts x vanish by line (1) of Eq. (1). Then, using Eq. (5), Section 2–1, with $\mathbf{K}$ replacing $\mathbf{L}$, and lines (2) and (3) of Eq. (1), the expression (6) is transformed into

$$\begin{aligned}(K_y K_x + iK_z)U_y + (K_z K_x - iK_y)U_z \\ - K_y(K_x U_y - iU_z) - K_z(K_x U_z + iU_y),\end{aligned} \tag{7}$$

which vanishes identically.

(b) *Matrix elements.*   We next compute matrix elements, in the function space defined by

$$\begin{aligned}\mathbf{K}^2\psi(\dots|\alpha Km) &= K(K+1)\psi(\dots|\alpha Km), \\ K_z\psi(\dots|\alpha Km) &= m\psi(\dots|\alpha Km),\end{aligned} \tag{8}$$

with the additional property that the only nonvanishing matrix elements of $\mathbf{K}$ are diagonal in the other quantum numbers α of the complete orthonormal set.   It follows, then, from Eqs. (11)–(13), Section 3–2, that these matrix elements are independent of α.

THEOREM II.   *The matrix elements of a scalar operator* V *vanish if* $K',m' \neq K,m$; *scalar here means that* V *is invariant under the rotations generated by* $\mathbf{K}$. *A particular case of importance is* $V = \mathbf{K} \cdot \mathbf{U}$.

Proof.   The relation $[\mathbf{K},V] = 0$ expresses the fact that V is invariant under a rigid rotation of the physical system.   This commutation rule implies a relation between matrix elements:

$$(\alpha'K'm|K_z V - VK_z|\alpha Km) = 0 \tag{9}$$

or

$$[(\alpha'K'm'|K_z|\alpha'K'm') - (\alpha Km|K_z|\alpha Km)](\alpha'K'm'|V|\alpha Km) = 0, \tag{10}$$

which is identical with

$$(m' - m)(\alpha'K'm'|V|\alpha Km) = 0. \tag{11}$$

Consequently,

$$(\alpha'K'm'|V|\alpha Km) = 0, \quad m' \neq m. \tag{12}$$

In a precisely similar way we can replace K_z by $\mathbf{K}^2$ in (9) above to show that matrix elements of V vanish if $K' \neq K$. We cannot assert, however, that α' and α must be identical for nonvanishing matrix elements of V.

THEOREM III.   *The matrix elements $(\alpha'Km|V|\alpha Km)$ are independent of $m(-K \leqq m \leqq K)$.*

Proof.   We list four results useful in establishing this theorem:

(a) $NV = VN$;
(b) $(\alpha Km + 1|N|\alpha Km) =$
$\sqrt{(K - m)(K + m + 1)}$;   [Eq. (9), of Section 3–2]
(c) all other matrix elements of N vanish;
(d) $(\alpha'Km + 1|NV - VN|\alpha Km) = 0$.

Introducing (b), (c), and (d) into the expansion theorem for the matrix elements of a product operator, we obtain

$$(\alpha Km + 1|N|\alpha Km)[(\alpha'Km|V|\alpha Km)$$
$$- (\alpha'Km + 1|V|\alpha Km + 1)] = 0. \tag{13}$$

Setting the term in square brackets equal to zero verifies that $(\alpha'Km|V|\alpha Km)$ is independent of m in the specified range.

Theorem IV, to follow, requires that we evaluate the second commutator of $\mathbf{K}^2$ and $\mathbf{U}$. We illustrate the procedures used in the derivation by finding the first commutator. Consider

$$\mathbf{K}^2 U_x - U_x \mathbf{K}^2 = (K_y^2 + K_z^2)U_x - U_x(K_y^2 + K_z^2). \tag{14}$$

The terms with subscripts x exclusively cancel by virtue of Eq. (1), line (1). Application of Eq. (1), line (6) to the commutator of K_y^2 and U_x yields

$$K_y^2 U_x - U_x K_y^2 = K_y(U_x K_y - iU_z) - (K_y U_x + iU_z)K_y$$
$$= -i(K_y U_z + U_z K_y). \tag{15}$$

In the same way,

$$K_z^2 U_x - U_x K_z^2 = i(K_z U_y + U_y K_z). \tag{16}$$

Equations (14), (15), and (16) combine to give

$$\mathbf{K}^2 \mathbf{U} - \mathbf{U} \mathbf{K}^2 = i(\mathbf{U} \times \mathbf{K} - \mathbf{K} \times \mathbf{U}). \tag{17}$$

The expanded form of the second commutator, $[\mathbf{K}^2, [\mathbf{K}^2, \mathbf{U}]]$, is

$$\mathbf{K}^4 \mathbf{U} - 2\mathbf{K}^2 \mathbf{U} \mathbf{K}^2 + \mathbf{U} \mathbf{K}^4.$$

The reader can verify with the help of numerous applications of Eq. (1) that

$$\mathbf{K}^4 \mathbf{U} - 2\mathbf{K}^2 \mathbf{U} \mathbf{K}^2 + \mathbf{U} \mathbf{K}^4 = 2(\mathbf{K}^2 \mathbf{U} + \mathbf{U} \mathbf{K}^2) - 4\mathbf{K}(\mathbf{K} \cdot \mathbf{U}). \tag{18}$$

THEOREM IV. $(\alpha' K m' | \mathbf{U} | \alpha K m) = B(\alpha', \alpha, K)(\alpha K m' | \mathbf{K} | \alpha K m).$ (19)

In classical terms, Eq. (19) states that (a) the component of $\mathbf{U}$ perpendicular to $\mathbf{K}$ has a vanishing time average because of the precession of $\mathbf{U}$ about $\mathbf{K}$, and (b) the projection of $\mathbf{U}$ on $\mathbf{K}$ is independent of the orientation of $\mathbf{K}$. A common application occurs when $\mathbf{U}$ is the magnetic moment operator; the factor $B(\alpha, \alpha', K)$ is then essentially a Landé g-factor.

Derivation. Equation (18) is converted into the matrix element relation

$$(\alpha' K' m' | \mathbf{K}^4 \mathbf{U} - 2\mathbf{K}^2 \mathbf{U} \mathbf{K}^2 + \mathbf{U} \mathbf{K}^4 - 2(\mathbf{K}^2 \mathbf{U} + \mathbf{U} \mathbf{K}^2) | \alpha K m)$$
$$= -4(\alpha' K' m' | \mathbf{K}(\mathbf{K} \cdot \mathbf{U}) | \alpha K m). \tag{20}$$

The Hermitian property of $\mathbf{K}^2$ and $\mathbf{U}$ allows the reduction of Eq. (20) to

$$(\alpha' K' m' | \mathbf{U} | \alpha K m)[K'^2(K' + 1)^2 - 2K'(K' + 1)K(K + 1)$$
$$+ K^2(K + 1)^2 - 2K'(K' + 1) - 2K(K + 1)]$$
$$= -4(\alpha' K' m' | \mathbf{K}(\mathbf{K} \cdot \mathbf{U}) | \alpha K m). \tag{21}$$

For $K' = K$, Eq. (21) reduces to

$$(\alpha' K m' | \mathbf{U} | \alpha K m) = \frac{(\alpha' K m' | \mathbf{K}(\mathbf{K} \cdot \mathbf{U}) | \alpha K m)}{K(K + 1)},$$

which can be further simplified with the help of the expansion theorem for the matrix elements of a product operator. We make use of the facts that $\mathbf{K}$ is diagonal in α and K (Section 3–2) and that $\mathbf{K} \cdot \mathbf{U}$

is diagonal in m (Theorem II). Since the matrix element of $\mathbf{K} \cdot \mathbf{U}$ is independent of m (Theorem III), it is convenient to denote $(\alpha' K m | \mathbf{K} \cdot \mathbf{U} | \alpha K m)$ by $(\alpha' K \| \mathbf{K} \cdot \mathbf{U} \| \alpha K)$. Finally, the matrix element of $\mathbf{U}$ is

$$(\alpha' K m' | \mathbf{U} | \alpha K m) = \frac{(\alpha' K \| \mathbf{K} \cdot \mathbf{U} \| \alpha K)}{K(K + 1)} (K m' | \mathbf{K} | K m). \qquad (22)$$

THEOREM V. *The only nonvanishing matrix elements of* $\mathbf{U}$ *have* $K' = K$ *or* $K' = K \pm 1$.

The first condition $K' = K$ is treated in the preceding theorem. The other possibilities for nonvanishing matrix elements can be derived from Eq. (21), assuming explicitly $K' \neq K$. This makes the right member vanish, since the matrix elements of $\mathbf{K}$ and also $\mathbf{K} \cdot \mathbf{U}$ are diagonal in K:

$$\begin{aligned}(\alpha' K' m' | \mathbf{K}(\mathbf{K} \cdot \mathbf{U}) | \alpha K m) &= (K' m' | \mathbf{K} | K' m)(\alpha' K' m | \mathbf{K} \cdot \mathbf{U} | \alpha K m) \\ &= 0 \text{ for } K' \neq K.\end{aligned} \qquad (23)$$

Equation (21) then reduces, on factoring, to

$$[(K' - K)^2 - 1][(K' + K + 1)^2 - 1](\alpha' K' m' | \mathbf{U} | \alpha K m) = 0. \qquad (24)$$

The first factor leads to possible nonvanishing elements of $\mathbf{U}$ if $K' = K \pm 1$, as stated in the theorem. The second factor appears to allow nonvanishing elements if $K' = K = 0$; however, if $K' \equiv K = 0$, all matrix elements of $\mathbf{K}$ vanish (Section 3–2), with the consequence that all matrix elements of $\mathbf{U}$ must also vanish; this is proved immediately from Eqs. (25) and (29) below, which are simply obtained from Eqs. (1).

THEOREM VI. *The matrix elements of* $(\alpha' K' m' | U_z | \alpha K m)$ *vanish unless* $m' = m$, *and the elements of* $(\alpha' K' m' | U_x \pm i U_y | \alpha K m)$ *vanish unless* $m' = m \pm 1$.

Proof. The commutation relations (1) imply that

$$\begin{aligned}K_z U_z &= U_z K_z, \\ K_z (U_x - i U_y) &= (U_x - i U_y)(K_z - 1), \\ K_z (U_x + i U_y) &= (U_x + i U_y)(K_z + 1).\end{aligned} \qquad (25)$$

The corresponding matrix element relations are

$$(m' - m)(\alpha'K'm'|U_z|\alpha Km) = 0,$$
$$(m' - m + 1)(\alpha'K'm'|U_x - iU_y|\alpha Km) = 0, \tag{26}$$
$$(m' - m - 1)(\alpha'K'm'|U_x + iU_y|\alpha Km) = 0,$$

from which the theorem follows directly.

THEOREM VII. *The following matrix element relation is valid:*

$$\frac{(\alpha'K'm+2|U_x+iU_y|\alpha Km+1)}{[(K'-m-1)(K'+m+2)]^{\frac{1}{2}}} = \frac{(\alpha'K'm+1|U_x+iU_y|\alpha Km)}{[(K-m)(K+m+1)]^{\frac{1}{2}}}. \tag{27}$$

Proof. Equation (1) implies that

$$N(U_x + iU_y) - (U_x + iU_y)N = 0, \tag{28}$$
$$M(U_x + iU_y) - (U_x + iU_y)M = -2U_z. \tag{29}$$

The matrix element relation

$$(\alpha'K'm+2|N(U_x+iU_y)|\alpha Km) = (\alpha'K'm+2|(U_x+iU_y)N|\alpha Km) \tag{30}$$

yields

$$(\alpha'K'm + 2|N|\alpha'K'm + 1)(\alpha'K'm + 1|U_x + iU_y|\alpha Km)$$
$$= (\alpha'K'm + 2|U_x + iU_y|\alpha Km + 1)(\alpha Km + 1|N|\alpha Km), \tag{31}$$

since the only nonvanishing elements of N and $U_x + iU_y$ are $(\alpha Km + 1|N|\alpha Km)$ [Eq. (12), Section 3–2] and $(\alpha'K'm + 1|U_x + iU_y|\alpha Km)$ (Theorem VI). Putting explicit values for the matrix elements of N into Eq. (31), we obtain Eq. (27). This relation has a number of important consequences.

In the special case $K' = K$, each member of Eq. (27) is independent of m; i.e., the equation has the form $f(m + 1) = f(m)$. There exist multiplying factors which make possible the extension of this statement to other conditions for nonvanishing matrix elements.

$K' = K + 1$. Equation (27) is multiplied by

$$(K - m)^{\frac{1}{2}}(K + m + 2)^{-\frac{1}{2}},$$

with the result

$$\frac{(\alpha'K + 1m + 2|U_x + iU_y|\alpha Km + 1)}{[(K + m + 2)(K + m + 3)]^{\frac{1}{2}}}$$
$$= \frac{(\alpha'K + 1m + 1|U_x + iU_y|\alpha Km)}{[(K + m + 1)(K + m + 2)]^{\frac{1}{2}}}, \tag{32}$$

which has the form $f(m + 1) = f(m)$.

$K' = K - 1$.   Equation (27) is multiplied by

$$(K + m + 1)^{\frac{1}{2}}(K - m - 1)^{-\frac{1}{2}},$$

with the result

$$\frac{(\alpha'K - 1m + 2|U_x + iU_y|\alpha Km + 1)}{[(K - m - 1)(K - m - 2)]^{\frac{1}{2}}}$$
$$= \frac{(\alpha'K - 1m + 1|U_x + iU_y|\alpha Km)}{[(K - m)(K - m - 1)]^{\frac{1}{2}}}, \quad (33)$$

once more in the form $f(m + 1) = f(m)$.

From these ratios independent of m, we obtain the nonvanishing matrix elements of $U_x + iU_y$:

$$
\begin{aligned}
(\alpha'Km &+ 1|U_x + iU_y|\alpha Km) \\
&= C(\alpha'K; \alpha K)[(K - m)(K + m + 1)]^{\frac{1}{2}}, \\
(\alpha'K &+ 1m + 1|U_x + iU_y|\alpha Km) \\
&= -C(\alpha'K + 1; \alpha K)[(K + m + 1)(K + m + 2)]^{\frac{1}{2}}, \quad (34) \\
(\alpha'K &- 1m + 1|U_x + iU_y|\alpha Km) \\
&= C(\alpha'K - 1; \alpha K)[(K - m)(K - m - 1)]^{\frac{1}{2}}.
\end{aligned}
$$

The new symbol $C(\alpha'K'; \alpha K)$ stands for the numbers defined by the ratios in Eqs. (27), (32), and (33).   The negative sign associated with $C(\alpha'K + 1; \alpha K)$ is required to ensure that $C(\alpha'K'; \alpha K)$ has the Hermitian property (to be proved).

The matrix elements of $U_x - iU_y$ can be derived from Eq. (34) by taking the complex conjugate of the left and right members.   From the Hermitian property of U_x and U_y and the assumed Hermitian property of $C(\alpha'K'; \alpha K)$ (still to be proved) we obtain the nonvanishing matrix elements:

$$
\begin{aligned}
(\alpha'Km|U_x &- iU_y|\alpha Km + 1) \\
&= C(\alpha'K; \alpha K)[(K - m)(K + m + 1)]^{\frac{1}{2}}, \\
(\alpha'K &- 1m|U_x - iU_y|\alpha Km + 1) \\
&= -C(\alpha'K - 1; \alpha K)[(K + m)(K + m + 1)]^{\frac{1}{2}}, \quad (35) \\
(\alpha'K &+ 1m|U_x - iU_y|\alpha Km + 1) \\
&= C(\alpha'K + 1; \alpha K)[(K - m)(K - m + 1)]^{\frac{1}{2}}.
\end{aligned}
$$

Equation (29) yields

$$2(\alpha'K'm|U_z|\alpha Km) = (\alpha'K'm|U_x+iU_y|\alpha Km-1)(\alpha Km-1|M|\alpha Km)$$
$$- (\alpha'K'm|M|\alpha'K'm+1)(\alpha'K'm+1|U_x+iU_y|\alpha Km). \quad (36)$$

Substitution of explicit values into the right member yields

$$
\begin{aligned}
(\alpha'Km|U_z|\alpha Km) &= C(\alpha'K;\alpha K)m, \\
(\alpha'K + 1m|U_z|\alpha Km) &= C(\alpha'K + 1;\alpha K)[(K + 1)^2 - m^2]^{\frac{1}{2}}, \\
(\alpha'K - 1m|U_z|\alpha Km) &= C(\alpha'K - 1;\alpha K)[K^2 - m^2]^{\frac{1}{2}}.
\end{aligned}
\quad (37)
$$

The Hermitian property of U_z ensures that

$$\overline{C(\alpha'K';\alpha K)} = C(\alpha K;\alpha'K'). \quad (38)$$

This is the Hermitian property secured by the introduction of a negative sign in Eq. (34). The reader should verify that the matrix elements of **U** are proportional to those of **K** for $K' = K$ (a direct check on Theorem IV).

4–2 Tensor relations and matrix elements. In the developments which follow it is convenient to express Eqs. (1) in the form

$$
\begin{aligned}
M(U_x + iU_y) - (U_x + iU_y)M &= -2U_z, \\
MU_z - U_zM &= U_x - iU_y, \\
M(U_x - iU_y) - (U_x - iU_y)M &= 0, \\
N(U_x + iU_y) - (U_x + iU_y)N &= 0, \\
NU_z - U_zN &= -(U_x + iU_y), \\
N(U_x - iU_y) - (U_x - iU_y)N &= 2U_z, \\
K_z(U_x \pm iU_y) - (U_x \pm iU_y)K_z &= \pm(U_x \pm iU_y).
\end{aligned}
\quad (1)
$$

Some of these have already been used in proving Theorems VI and VII. The definitions

$$
\begin{aligned}
Y_{11} &= -(U_x + iU_y), \\
Y_{10} &= \sqrt{2}U_z, \\
Y_{1,-1} &= U_x - iU_y
\end{aligned}
\quad (2)
$$

permit writing Eq. (1) in a shorter, more elegant form:

$$
\begin{aligned}
MY_{1\mu+1} - Y_{1\mu+1}M &= \sqrt{(1 - \mu)(2 + \mu)}\,Y_{1\mu}, \\
NY_{1\mu} - Y_{1\mu}N &= \sqrt{(1 - \mu)(2 + \mu)}\,Y_{1\mu+1}, \\
K_zY_{1\mu} - Y_{1\mu}K_z &= \mu Y_{1\mu}.
\end{aligned}
\quad (3)
$$

Next we define tensor quantities in analogy to surface spherical harmonics of the second order:

$$
\begin{aligned}
Z_2 &= (U_x + iU_y)^2, \\
Z_1 &= -U_z(U_x + iU_y) - (U_x + iU_y)U_z, \\
Z_0 &= 3U_z^2 - \mathbf{U}^2, \\
Z_{-1} &= U_z(U_x - iU_y) + (U_x - iU_y)U_z, \\
Z_{-2} &= (U_x - iU_y)^2.
\end{aligned}
\tag{4}
$$

These functions satisfy the following commutation relations:

$$
\left.
\begin{aligned}
MZ_2 - Z_2M &= 2Z_1 \\
MZ_1 - Z_1M &= 2Z_0 \\
MZ_0 - Z_0M &= 3Z_{-1} \\
MZ_{-1} - Z_{-1}M &= 2Z_{-2} \\
MZ_{-2} - Z_{-2}M &= 0
\end{aligned}
\right\}
$$

$$
\left.
\begin{aligned}
NZ_2 - Z_2N &= 0 \\
NZ_1 - Z_1N &= 2Z_2 \\
NZ_0 - Z_0N &= 3Z_1 \\
NZ_{-1} - Z_{-1}N &= 2Z_0 \\
NZ_{-2} - Z_{-2}N &= 2Z_{-1}
\end{aligned}
\right\}
\tag{5}
$$

$$
K_zZ_\mu - Z_\mu K_z = \mu Z_\mu]
$$

The proof of Eq. (5) is omitted, since only obvious, but tedious, repeated applications of Eqs. (1) or (3) are involved.

A slight modification in the definition of the Z_μ functions enables us to express Eq. (5) in a form analogous to Eq. (3). We replace Z_μ by

$$
\begin{aligned}
Y_{22} &= (U_x + iU_y)^2, \\
Y_{21} &= -U_z(U_x + iU_y) - (U_x + iU_y)U_z, \\
Y_{20} &= \sqrt{\tfrac{2}{3}}(3U_z^2 - \mathbf{U}^2), \\
Y_{2,-1} &= U_z(U_x - iU_y) + (U_x - iU_y)U_z, \\
Y_{2,-2} &= (U_x - iU_y)^2,
\end{aligned}
\tag{6}
$$

and obtain from Eqs. (4) and (5):

$$
\begin{aligned}
MY_{2,\mu+1} - Y_{2,\mu+1}M &= \sqrt{(2-\mu)(3+\mu)}\,Y_{2,\mu}, \\
NY_{2,\mu} - Y_{2,\mu}N &= \sqrt{(2-\mu)(3+\mu)}\,Y_{2,\mu+1}, \\
K_zY_{2,\mu} - Y_{2,\mu}K_z &= \mu Y_{2,\mu}.
\end{aligned}
\tag{7}
$$

THEOREM VIII. *The third line of Eq. (7), when expressed as a relation between matrix elements, yields*

$$(\alpha' K' m' | Y_{2\mu} | \alpha K m) = 0, \tag{8}$$

unless

$$m' = m + \mu.$$

THEOREM IX. *Recurrence relations for matrix elements of* $Y_{2,\mu}$.

The starting point is the homogeneous relation $N Y_{22} = Y_{22} N$. In matrix element form, this yields

$$\frac{(\alpha' K' m + 2 | Y_{22} | \alpha K m)}{[(K' - m - 1)(K' + m + 2)]^{\frac{1}{2}}} = \frac{(\alpha' K' m + 1 | Y_{22} | \alpha K m - 1)}{[(K - m + 1)(K + m)]^{\frac{1}{2}}}. \tag{9}$$

The commutators $[M, Y_{2,\mu+1}]$ generate a relation between the matrix elements of $Y_{2,\mu+1}$ and $Y_{2,\mu}$ in the form

$$[(2 - \mu)(3 + \mu)]^{\frac{1}{2}} (\alpha' K' m + \mu | Y_{2,\mu} | \alpha K m)$$
$$= [(K' - m - \mu)(K' + m + \mu + 1)]^{\frac{1}{2}} (\alpha' K' m + \mu + 1 | Y_{2,\mu+1} | \alpha K m) \tag{10}$$
$$- [(K - m + 1)(K + m)]^{\frac{1}{2}} (\alpha' K' m + \mu | Y_{2,\mu+1} | \alpha K m - 1)$$

for $\mu = 1, 0, -1, -2$. The derivation of Eq. (10) from the appropriate commutation rule in Eq. (7) follows the familiar pattern employing the special properties of the M and N matrix elements in conjunction with the expansion theorem for the matrix elements of a product operator.

Case 1. $K' = K$.

Equation (9) can be expressed in the form $g(m) = g(m + 1)$ by dividing both terms by $[(K - m)(K + m + 1)]^{\frac{1}{2}}$. The resulting equation is equivalent to

$$\boxed{\begin{aligned} (\alpha' K m &+ 2 | Y_{22} | \alpha K m) \\ &= C[(K - m)(K - m - 1)(K + m + 1)(K + m + 2)]^{\frac{1}{2}}, \end{aligned}} \tag{11}$$

in which C is independent of m. Equation (10) is now used as a descending ladder of relations to compute matrix elements of Y_{21}, Y_{20}, $Y_{2,-1}$, $Y_{2,-2}$ in succession. The results are

$$
\begin{aligned}
(\alpha'Km &+ 1|Y_{21}|\alpha Km) \\
&= -C(2m + 1)[(K - m)(K + m + 1)]^{\frac{1}{2}}, \\
(\alpha'Km|Y_{20}&|\alpha Km) = C[3m^2 - K(K + 1)](\tfrac{2}{3})^{\frac{1}{2}}, \\
(\alpha'Km|Y_{2,-1}&|\alpha Km + 1) \\
&= C(2m + 1)[(K - m)(K + m + 1)]^{\frac{1}{2}}, \\
(\alpha'Km|Y_{2,-2}&|\alpha Km + 2) \\
&= C[(K - m)(K - m - 1)(K + m + 1)(K + m + 2)]^{\frac{1}{2}}.
\end{aligned}
\tag{12}
$$

These matrix elements all vanish for $K = 0$ or $\frac{1}{2}$.

The Hermitian property

$$
\overline{C(\alpha'K; \alpha K)} = C(\alpha K; \alpha'K)
\tag{13}
$$

is required by the connection between C and the nonvanishing matrix element of Y_{20}. Explicitly,

$$
C(\alpha'K; \alpha K) = (\tfrac{3}{2})^{\frac{1}{2}} \frac{(\alpha'KK|Y_{20}|\alpha KK)}{K(2K - 1)}.
\tag{14}
$$

For other values of K', division by the factors

$$
\begin{aligned}
[(K + m + 1)(K + m + 2)]^{\frac{1}{2}}, && K' &= K + 1, \\
[(K - m)(K - m - 1)]^{\frac{1}{2}}, && K' &= K - 1, \\
[(K + m + 1)(K + m + 2)(K + m + 3)]^{\frac{1}{2}}, && K' &= K + 2, \\
[(K - m)(K - m - 1)(K - m - 2)]^{\frac{1}{2}}, && K' &= K - 2
\end{aligned}
\tag{15}
$$

accomplishes the transformation of Eq. (9) into relations of the type $g(m + 1) = g(m)$. The remaining analysis, in particular the application of Eq. (10) for $\mu = 1, 0, -1$ and -2, is left as an exercise for the reader and results are simply stated:

Case 2. $K' = K + 1$.

$$
\begin{aligned}
(\alpha'K &+ 1m + 2|Y_{22}|\alpha Km) \\
&= C[(K - m)(K + m + 1)(K + m + 2)(K + m + 3)]^{\frac{1}{2}}, \\
(\alpha'K &+ 1m + 1|Y_{21}|\alpha Km) \\
&= C(K - 2m)[(K + m + 1)(K + m + 2)]^{\frac{1}{2}}, \\
(\alpha'K &+ 1m|Y_{20}|\alpha Km) \\
&= -Cm[6(K - m + 1)(K + m + 1)]^{\frac{1}{2}}, \\
(\alpha'K &+ 1m|Y_{2,-1}|\alpha Km + 1) \\
&= -C(K + 2m + 2)[(K - m + 1)(K - m)]^{\frac{1}{2}}, \\
(\alpha'K &+ 1m|Y_{2,-2}|\alpha Km + 2) \\
&= -C[(K - m + 1)(K - m)(K - m - 1)(K + m + 2)]^{\frac{1}{2}}.
\end{aligned}
\tag{16}
$$

$$\overline{C(\alpha'K + 1; \alpha K)} = C(\alpha K; \alpha'K + 1).$$

These matrix elements all vanish for $K = 0$.

Case 3. $K' = K - 1$.

$$
\begin{aligned}
&(\alpha'K - 1m + 2| Y_{22}|\alpha Km) \\
&= -C[(K - m)(K - m - 1)(K - m - 2)(K + m + 1)]^{\frac{1}{2}}, \\
&(\alpha'K - 1m + 1| Y_{21}|\alpha Km) \\
&\qquad = C(K + 2m + 1)[(K - m)(K - m - 1)]^{\frac{1}{2}}, \\
&(\alpha'K - 1m| Y_{20}|\alpha Km) = -Cm[6(K - m)(K + m)]^{\frac{1}{2}}, \\
&(\alpha'K - 1m| Y_{2,-1}|\alpha Km + 1) \\
&\qquad = -C(K - 2m - 1)[(K + m)(K + m + 1)]^{\frac{1}{2}}, \\
&(\alpha'K - 1m| Y_{2,-2}|\alpha Km + 2) \\
&= C[(K - m - 1)(K + m)(K + m + 1)(K + m + 2)]^{\frac{1}{2}}.
\end{aligned}
\tag{17}
$$

$$\overline{C(\alpha'K - 1; \alpha K)} = C(\alpha K; \alpha'K - 1).$$

These matrix elements all vanish for $K = 1$.

Case 4. $K' = K + 2$.

$$
\begin{aligned}
&(\alpha'K + 2m + 2| Y_{22}|\alpha Km) \\
&= C[(K + m + 1)(K + m + 2)(K + m + 3)(K + m + 4)]^{\frac{1}{2}}, \\
&(\alpha'K + 2m + 1| Y_{21}|\alpha Km) \\
&= 2C[(K - m + 1)(K + m + 1)(K + m + 2)(K + m + 3)]^{\frac{1}{2}}, \\
&(\alpha'K + 2m| Y_{20}|\alpha Km) \\
&= C[6(K - m + 2)(K - m + 1)(K + m + 1)(K + m + 2)]^{\frac{1}{2}}, \\
&(\alpha'K + 2m| Y_{2,-1}|\alpha Km + 1) \\
&= 2C[(K - m + 2)(K - m + 1)(K - m)(K + m + 2)]^{\frac{1}{2}}, \\
&(\alpha'K + 2m| Y_{2,-2}|\alpha Km + 2) \\
&= C[(K - m + 2)(K - m + 1)(K - m)(K - m - 1)]^{\frac{1}{2}}.
\end{aligned}
\tag{18}
$$

$$\overline{C(\alpha'K + 2; \alpha K)} = C(\alpha K; \alpha'K + 2).$$

Case 5. $K' = K - 2.$

$$
\begin{aligned}
&(\alpha'K - 2m + 2|\,Y_{22}|\alpha Km) \\
&\quad = C[(K - m)(K - m - 1)(K - m - 2)(K - m - 3)]^{\frac{1}{2}}, \\
&(\alpha'K - 2m + 1|\,Y_{21}|\alpha Km) \\
&\quad = -2C[(K + m)(K - m)(K - m - 1)(K - m - 2)]^{\frac{1}{2}}, \\
&(\alpha'K - 2m|\,Y_{20}|\alpha Km) \\
&\quad = C[6(K + m - 1)(K + m)(K - m)(K - m - 1)]^{\frac{1}{2}}, \\
&(\alpha'K - 2m|\,Y_{2,-1}|\alpha Km + 1) \\
&\quad = -2C[(K - m - 1)(K + m - 1)(K + m)(K + m + 1)]^{\frac{1}{2}}, \\
&(\alpha'K - 2m|\,Y_{2,-2}|\alpha Km + 2) \\
&\quad = C[(K + m - 1)(K + m)(K + m + 1)(K + m + 2)]^{\frac{1}{2}}.
\end{aligned} \tag{19}
$$

$$
\overline{C(\alpha'K - 2; \alpha K)} = C(\alpha K; \alpha'K - 2).
$$

Equations (3) and (7) express the fact that quantities of the form $Y_{1\mu}(\mathbf{U})$ and $Y_{2\mu}(\mathbf{U})$ transform like eigenfunctions of the angular momentum operators $\mathbf{K}^2, K_z$ with the eigenvalues $1(1 + 1), \mu$ and $2(2 + 1), \mu$ respectively. The definition

$$
Y_{KK}(\mathbf{U}) = (-1)^K (U_x + iU_y)^K \tag{20}
$$

and the relation

$$
MY_{K\mu+1} - Y_{K\mu+1}M = \sqrt{(K - \mu)(K + \mu + 1)}\,Y_{K\mu} \tag{21}
$$

define a set of tensor operators $Y_{K\mu}$ which transform like eigenfunctions of $\mathbf{K}^2, K_z$ with the eigenvalues $K(K + 1), \mu$. For the further development of this theme, the reader should consult the references at the end of Chapter 5.

CHAPTER 5

APPLICATIONS TO NUCLEAR MOMENTS AND TRANSITION PROBABILITIES

The preceding theorems may be applied to calculate energy levels of nuclei with magnetic dipole and electric quadrupole moments. No explicit reference to either the elementary vector model of the nuclear magnet or to group theoretical methods is needed. Only a few basic formulas are derived to illustrate the use of the theorems; consideration of special nuclear models falls outside the scope of these notes.

5-1 Nuclear magnetic interactions with a uniform magnetic field. The classical Hamiltonian for a particle of charge e and mass M in an electromagnetic field is

$$\mathcal{3C} = \frac{1}{2M}\left(\mathbf{p} - \frac{e}{c}\mathbf{A}\right)^2 + e\phi, \tag{1}$$

where $\mathbf{p}$ is the linear momentum vector and $\mathbf{A}$ and ϕ are the vector and scalar potentials.

In the special case of a constant magnetic field H_0 directed along the positive z-axis, the components of $\mathbf{A}$ may be written

$$A_x = -\tfrac{1}{2}yH_0, \quad A_y = \tfrac{1}{2}xH_0, \quad A_z = 0. \tag{2}$$

With neglect of second order terms in H_0, the expanded Hamiltonian is

$$\begin{aligned}
\mathcal{3C} &= \frac{1}{2M}\mathbf{p}^2 + e\phi - \frac{eH_0}{2Mc}(xp_y - yp_x) \\
&= \frac{1}{2M}\mathbf{p}^2 + e\phi - \frac{e\hbar H_0}{2Mc}L_z.
\end{aligned} \tag{3}$$

For arbitrary orientation of the field,

$$\mathcal{3C} = \frac{1}{2M}\mathbf{p}^2 + e\phi - \frac{e\hbar}{2Mc}\mathbf{L}\cdot\mathbf{H}_0. \tag{4}$$

Comparison with the classical potential energy $-\boldsymbol{\mu}\cdot\mathbf{H}_0$ of a magnetic dipole $\boldsymbol{\mu}$ in a magnetic field yields

43

$$\mathbf{\mu} = \frac{e\hbar}{2Mc}\, \mathbf{L}\cdot \tag{5}$$

for the orbital magnetic moment of the charged particle.

In addition, magnetic moments are associated with the intrinsic angular momentum of the neutrons and protons in the nucleus. The total nucleonic moment operator is therefore

$$\mathbf{\mu} = \frac{e\hbar}{2Mc}\, (\mathbf{L}_p + g_{sp}\mathbf{S}_p + g_{sn}\mathbf{S}_n), \tag{6}$$

in which the gyromagnetic factors g_{sp} and g_{sn} as found for isolated nucleons are

$$g_{sp} = 5.5850, \quad g_{sn} = -3.8206.$$

In this expression, subscripts p and n refer respectively to protons and neutrons. The orbital angular momentum operator of the neutrons, $\mathbf{L}_n$, does not appear, because the motion of an uncharged particle cannot generate an orbital magnetic moment.

The matrix elements of $\mathbf{\mu}$ are related to those of $\mathbf{I}$ through Eq. (22) of Section 4–1, in which $\mathbf{\mu}$ is identified with $\mathbf{U}$ and $\mathbf{I}$ with $\mathbf{K}$. Thus

$$(\alpha I m' |\mathbf{\mu}| \alpha I m) = g\, \frac{e\hbar}{2Mc}\, (I m' |\mathbf{I}| I m), \tag{7}$$

where g is the Landé g-factor of the state αI. The constant factor $e\hbar/2Mc$ is conveniently denoted by the symbol μ_0. Under the name of nuclear magneton it provides a convenient unit in which nuclear moments are usually expressed.

Equation (7) yields an effective moment operator

$$\mathbf{\mu} = g\mu_0\mathbf{I} \tag{8}$$

to replace the exact definition of Eq. (6). The effective component of $\mathbf{\mu}$ along the z-axis is therefore $g\mu_0 m$ for the state $\alpha I m$. For $m = I$, $\mu_z = g\mu_0 I$ is defined to be the magnetic moment μ. The dimensionless number gI is usually listed in tabulations as the magnetic moment of a nucleus.

These considerations yield a magnetic dipole Hamiltonian

$$\mathcal{3C}_\mu = -\mathbf{\mu} \cdot \mathbf{H}_0 = -g\mu_0\mathbf{I} \cdot \mathbf{H}_0 \tag{9}$$

with diagonal matrix elements

$$E_{\mu m} = (\alpha I m | \mathfrak{K}_\mu | \alpha I m) = -g\mu_0 H_0 m \qquad (10)$$

forming a ladder of $2I + 1$ energy levels spaced uniformly $g\mu_0 H_0$ apart.

5–2 Electric quadrupole moment. Consider the interaction of the nuclear protons with a fixed external charge distribution producing a static potential $V(\mathbf{r})$. The electrostatic potential energy of the ith proton is then $V(\mathbf{r}_i)$, with the origin taken for convenience at the centroid of the nuclear charge. The electrostatic energy operator for the nucleus is

$$\mathfrak{K}_e = \sum_i eV(\mathbf{r}_i)$$
$$= e\sum_i \left[V(0) + \sum_{\mu=1}^3 \left(\frac{\partial V}{\partial x_\mu}\right)_0 x_{\mu i} + \frac{1}{2}\sum_{\mu,\nu}\left(\frac{\partial^2 V}{\partial x_\mu\,\partial x_\nu}\right)_0 x_{\mu i}x_{\nu i} + \cdots \right], \qquad (1)$$

where e is the proton charge and $x_{\mu i}$ ranges over the x,y,z coordinates of all the protons. To find the energy shifts for a nuclear system in the state αI we require the matrix elements $(\alpha I m' | \mathfrak{K}_e | \alpha I m)$.

The constant term $V(0)$ requires no discussion, since it is equally effective in shifting all levels.

Since terms in x_k have odd parity, they contribute nothing to matrix elements diagonal in αI; an equivalent statement is that a nucleus cannot possess a static electric dipole moment of electrostatic origin. This follows from the fact that no physical basis exists for a difference in behavior of an isolated nuclear system in the two spatial configurations $\mathbf{r}_1, \mathbf{r}_2, \ldots$ and $-\mathbf{r}_1, -\mathbf{r}_2, \ldots$. Consequently, a parity quantum number exists and a wave function $\psi(\ldots | \alpha I m)$ either reverses sign or remains the same in going from one configuration to the other. It is also clear that the parity of $\psi(\ldots | \alpha I m)$ is independent of m, since the displacement operators are invariant under reflection through the origin. Consequently,

$$\bar{\psi}(\ldots \mathbf{r}_i \ldots | \alpha I m')\psi(\ldots \mathbf{r}_i \ldots | \alpha I m)$$
$$= \bar{\psi}(\ldots -\mathbf{r}_i \ldots | \alpha I m')\psi(\ldots -\mathbf{r}_i \ldots | \alpha I m); \qquad (2)$$

the product has even parity.

The second order term of Eq. (1) is the quadrupole interaction of interest in the present discussion. The matrix elements of the quadrupole Hamiltonian

$$\mathcal{H}_Q = \frac{1}{2} \sum_{i\mu\nu} e \left(\frac{\partial^2 V}{\partial x_\mu \, \partial x_\nu} \right)_0 x_{\mu i} x_{\nu i} \tag{3}$$

are easily computed from the theorems of Chapter 4 with aid of an identity:

$$\begin{aligned}
\sum_{\mu\nu} a_{\mu\nu} x_\mu x_\nu = {} & \tfrac{1}{4}(a_{11} - a_{22} - 2ia_{12})(x + iy)^2 \\
& + \tfrac{1}{4}(a_{11} - a_{22} + 2ia_{12})(x - iy)^2 \\
& + (a_{13} - ia_{23})z(x + iy) \\
& + (a_{13} + ia_{23})z(x - iy) \\
& + \tfrac{1}{6}(2a_{33} - a_{11} - a_{22})(2z^2 - x^2 - y^2) \\
& + \tfrac{1}{3}(a_{11} + a_{22} + a_{33})(x^2 + y^2 + z^2).
\end{aligned} \tag{4}$$

To illustrate the procedure, we carry through the calculation for $m' = m$:

$$\begin{aligned}
(\alpha Im | \mathcal{H}_Q | \alpha Im) = {} & \frac{e}{12} \left[2\left(\frac{\partial^2 V}{\partial z^2} \right)_0 - \left(\frac{\partial^2 V}{\partial x^2} \right)_0 - \left(\frac{\partial^2 V}{\partial y^2} \right)_0 \right] \\
& \cdot \left(\alpha Im \left| \sum_i (2z_i^2 - x_i^2 - y_i^2) \right| \alpha Im \right).
\end{aligned} \tag{5}$$

Laplace's equation holds for $V(\mathbf{r})$ in the region of the nucleus; consequently

$$2\left(\frac{\partial^2 V}{\partial z^2} \right)_0 - \left(\frac{\partial^2 V}{\partial x^2} \right)_0 - \left(\frac{\partial^2 V}{\partial y^2} \right)_0 = 3\left(\frac{\partial^2 V}{\partial z^2} \right)_0 \tag{6}$$

and

$$(\alpha Im | \mathcal{H}_Q | \alpha Im) = \frac{1}{4} e \left(\frac{\partial^2 V}{\partial z^2} \right)_0 \left(\alpha Im \left| \sum_i (3z_i^2 - r_i^2) \right| \alpha Im \right). \tag{7}$$

The last form can now be related to the matrix elements of $3I_z^2 - \mathbf{I}^2$ [Eq. (12) of Section 4–2]. Writing

$$Q = \left(\alpha II \left| \sum_i (3z_i^2 - r_i^2) \right| \alpha II \right), \tag{8}$$

we get

$$(\alpha Im | \mathcal{H}_Q | \alpha Im) = \frac{1}{4} e \left(\frac{\partial^2 V}{\partial z^2} \right)_0 \frac{Q}{I(2I - 1)} \{3m^2 - I(I + 1)\}. \tag{9}$$

The quantity Q is called the *quadrupole moment*. Since Q vanishes for a spherically symmetric distribution of charge, it provides a useful measure of the deviation from spherical symmetry of the nuclear charge distribution. Positive Q signifies elongation of the charge distribution along the symmetry axis (cigar or football-like shape). Negative Q signifies a distribution flattened at the poles and bulging at the equator.

5–3 Quadrupole interaction between atomic electrons and nuclei. The nuclear quadrupole moment appears in the electrostatic interaction between atomic electrons and nuclear protons. Taking account only of the field produced by electrons outside the nucleus, the expansion

$$\frac{1}{|\mathbf{r}_e - \mathbf{r}_p|} - \frac{1}{|\mathbf{r}_e|} = \frac{\mathbf{r}_e \cdot \mathbf{r}_p}{r_e^3} + \frac{3(\mathbf{r}_e \cdot \mathbf{r}_p)^2 - r_e^2 r_p^2}{2r_e^5} + \cdots \qquad (1)$$

exhibits the perturbing energy as a sum of dipole-dipole and quadrupole-quadrupole terms. As shown in the preceding section, all the nuclear wave functions in the αI subspace have the same parity; consequently the nucleus does not possess a permanent electric dipole moment and the dipole-dipole terms produce no first-order displacement of the energy levels.

Two identities are useful in reducing the first-order quadrupole-quadrupole energy to diagonal form:

$$\begin{aligned}
3(\mathbf{r}_e \cdot \mathbf{r}_p)^2 - r_e^2 r_p^2 &= \tfrac{1}{2}(3z_e^2 - r_e^2)(3z_p^2 - r_p^2) \\
&+ 3z_e(x_e + iy_e)z_p(x_p - iy_p) \\
&+ 3z_e(x_e - iy_e)z_p(x_p + iy_p) \\
&+ \tfrac{3}{4}(x_e + iy_e)^2(x_p - iy_p)^2 \\
&+ \tfrac{3}{4}(x_e - iy_e)^2(x_p + iy_p)^2.
\end{aligned} \qquad (2)$$

$$\begin{aligned}
3(\mathbf{I} \cdot \mathbf{J})^2 - \mathbf{I}^2\mathbf{J}^2 + \tfrac{3}{2}\mathbf{I} \cdot \mathbf{J} &= \tfrac{1}{2}(3J_z^2 - J^2)(3I_z^2 - I^2) \\
&+ \tfrac{3}{4}[J_z(J_x+iJ_y)+(J_x+iJ_y)J_z][I_z(I_x-iI_y)+(I_x-iI_y)I_z] \\
&+ \tfrac{3}{4}[J_z(J_x-iJ_y)+(J_x-iJ_y)J_z][I_z(I_x+iI_y)+(I_x+iI_y)I_z] \\
&+ \tfrac{3}{4}(J_x+iJ_y)^2(I_x-iI_y)^2 + \tfrac{3}{4}(J_x-iJ_y)^2(I_x+iI_y)^2.
\end{aligned} \qquad (3)$$

Equations (2) and (3) differ in form because in the first all quantities commute, while in the second the characteristic failure of commutativity for the components of angular momentum vectors gives rise to additional terms. These identities and a double application of

Eqs. (11) and (12) of Section 4–2 reduce the quadrupole-quadrupole matrix elements in the $\alpha I \beta J$ function space to

$$\left(\beta J m'_J \alpha I m'_I \middle| \sum_{e,p} \{3(\mathbf{r}_e \cdot \mathbf{r}_p)^2 - r_e^2 r_p^2\}/r_e^5 \middle| \beta J m_J \alpha I m_I \right)$$
$$= C(\beta J) C(\alpha I)(\beta J m'_J \alpha I m'_I | 3(\mathbf{I} \cdot \mathbf{J})^2 + \tfrac{3}{2}\mathbf{I} \cdot \mathbf{J} - \mathbf{I}^2 \mathbf{J}^2 | \beta J m_J \alpha I m_I), \tag{4}$$

in which

$$C(\beta J) = \frac{1}{J(2J-1)} \left(\beta J J \middle| \sum_e (3z_e^2 - r_e^2)/r_e^5 \middle| \beta J J \right), \tag{5}$$

$$C(\alpha I) = \frac{1}{I(2I-1)} \left(\alpha I I \middle| \sum_p (3z_p^2 - r_p^2) \middle| \beta I I \right)$$
$$= \frac{Q}{I(2I-1)}. \tag{6}$$

Now we transform to the representation in which the total angular momentum operators $\mathbf{F}^2 = (\mathbf{I} + \mathbf{J})^2$ and $F_z = I_z + J_z$ have definite numerical values, $F(F+1)$ and m_F. In this representation, $2\mathbf{I} \cdot \mathbf{J}$ has the definite numerical value $F(F+1) - J(J+1) - I(I+1)$ (denoted by K for brevity). Consequently,

$$3(\mathbf{I} \cdot \mathbf{J})^2 + \tfrac{3}{2}\mathbf{I} \cdot \mathbf{J} - \mathbf{I}^2 \mathbf{J}^2 \rightarrow \tfrac{3}{4}K(K+1) - I(I+1)J(J+1), \tag{7}$$

and the eigenvalues of the quadrupole-quadrupole energy matrix in the $\beta J \alpha I$ subspace are

$$\Delta E(\beta J \alpha I; \beta \alpha F) = e^2 C(\beta J) C(\alpha I) \cdot [\tfrac{3}{4}K(K+1) - I(I+1)J(J+1)]. \tag{8}$$

5–4 Landé g-factor for a two-component system. Operators of the type

$$\boldsymbol{\mu} = g_1 \mathbf{I}_1 + g_2 \mathbf{I}_2 \tag{1}$$

occur in many physical problems. We evaluate the diagonal matrix elements of μ_z in a representation labeled with eigenvalues of $\mathbf{I}_1^2, \mathbf{I}_2^2$, $\mathbf{I}^2 = (\mathbf{I}_1 + \mathbf{I}_2)^2, I_{1z}, I_{2z}$, and $I_z = I_{1z} + I_{2z}$.

Theorem IV of Chapter 4 yields

$$(\alpha I m | \mu_z | \alpha I m) = (I m | I_z | I m) \frac{(\alpha I | \mathbf{I} \cdot \boldsymbol{\mu} | \alpha I)}{I(I+1)}$$
$$= \frac{m}{I(I+1)} [g_1(\alpha I | \mathbf{I} \cdot \mathbf{I}_1 | \alpha I) + g_2(\alpha I | \mathbf{I} \cdot \mathbf{I}_2 | \alpha I)]. \tag{2}$$

From $I_2 = I - I_1$ and $I_1 = I - I_2$, we get the result that the wave functions are eigenfunctions of $\mathbf{I} \cdot \mathbf{I}_1$ and $\mathbf{I} \cdot \mathbf{I}_2$, with the eigenvalues

$$2\mathbf{I} \cdot \mathbf{I}_1 \to I(I + 1) + I_1(I_1 + 1) - I_2(I_2 + 1),$$
$$2\mathbf{I} \cdot \mathbf{I}_2 \to I(I + 1) - I_1(I_1 + 1) + I_2(I_2 + 1). \tag{3}$$

Consequently,

$$(\alpha I m | \mu_z | \alpha I m) = m \left[\tfrac{1}{2}(g_1 + g_2) + \tfrac{1}{2}(g_1 - g_2) \frac{I_1(I_1 + 1) - I_2(I_2 + 1)}{I(I + 1)} \right] \tag{4}$$

If μ is a magnetic moment operator, the factor in square brackets is known as the Landé g-factor of the αI state.

5–5 Equivalent operators in LS and jj coupling states. In the state with eigenvalues

$$\mathbf{L}^2 \to L(L + 1),$$
$$\mathbf{S}^2 \to S(S + 1) = \tfrac{1}{2}(\tfrac{1}{2} + 1), \tag{1}$$

the total angular momentum can take on two different values: $I = L + \tfrac{1}{2}$ without restriction on L, and $I = L - \tfrac{1}{2}$ for $L > 0$. The operator $2\mathbf{I} \cdot \mathbf{S}$ has the eigenvalues $I + 1$ ($I = L + \tfrac{1}{2}$) and $-I$ ($I = L - \tfrac{1}{2}$). These results combine with Theorem IV of Chapter 4 to yield

$$(\alpha I m' | \mathbf{S} | \alpha I m) = \frac{(\alpha I m' | \mathbf{I} | \alpha I m)}{2L + 1} (-1)^{L + \frac{1}{2} - I}. \tag{2}$$

In the spin-orbit coupling shell model, the single particle states have $L = l$ and $I = j$. Equation (2) then expresses the equivalence

$$\mathbf{S} \sim \frac{\mathbf{j}}{2l + 1} (-1)^{l + \frac{1}{2} - j} \tag{3}$$

for the single particle operators in the αj function space.

An interesting application of Eq. (3) occurs in evaluating the diagonal matrix elements of $\mathbf{S}^2$ for a state generated by n particles in equivalent j orbitals:

$$\mathbf{S}^2 = \sum_{p,q} \mathbf{S}_p \cdot \mathbf{S}_q$$
$$\sim \frac{3}{4} n + \sum_{p \neq q} \frac{\mathbf{j}_p \cdot \mathbf{j}_q}{(2l + 1)^2} \tag{4}$$
$$\to n \left[\frac{3}{4} - \frac{j(j + 1)}{(2l + 1)^2} \right] + \frac{I(I + 1)}{(2l + 1)^2}.$$

This result is correct for a $j = l \pm \frac{1}{2}$ subshell only if the complementary subshell with $j' = l \mp \frac{1}{2}$ is empty. A cross term occurs when orbitals are occupied in both subshells. The cross terms can be computed from the condition that the mean value of S^2 vanishes when both j subshells are completely filled. In particular, for a state having a filled $l + \frac{1}{2}$ subshell and n' holes in the $j = l - \frac{1}{2}$ subshell,

$$S^2 \to n' \left[\frac{3}{4} - \frac{j(j+1)}{(2l+1)^2} \right] + \frac{I(I+1)}{(2l+1)^2}. \tag{5}$$

5–6 Intensity relations. Many problems require evaluation of a quantity

$$\sum_{m'} |(\alpha'I'm'|\mathbf{U}|\alpha Im)|^2 = \frac{1}{2} \sum_{m'\mu} |(\alpha'I'm'|Y_{1\mu}|\alpha Im)|^2. \tag{1}$$

The explicit representation of the component matrix elements given in Eqs. (34), (35), and (37) of Section 4–1 yields the following theorems:

THEOREM X.

$$\frac{1}{2} \sum_{\mu m'} |(\alpha'I + 1m'|Y_{1\mu}|\alpha Im)|^2$$
$$= \frac{(I+1)(2I+3)}{2I+1} |(\alpha'I + 1I|U_z|\alpha II)|^2, \tag{2}$$

$$\frac{1}{2} \sum_{\mu m'} |(\alpha'Im'|Y_{1\mu}|\alpha Im)|^2 = \frac{I+1}{I} |(\alpha'II|U_z|\alpha II)|^2, \tag{3}$$

$$\frac{1}{2} \sum_{\mu m'} |(\alpha'I - 1m'|Y_{1\mu}|\alpha Im)|^2 = I|(\alpha'I - 1, I - 1|U_z|\alpha I, I - 1)|^2. \tag{4}$$

Equation (3) excludes $I = 0$; all matrix elements of $Y_{1\mu}$ vanish in this case.

The following physical statement is an immediate consequence of Theorem X: the transition probability associated with the operator $\mathbf{U}$ from an initial state αIm to all magnetic substates of $\alpha'I'm'$ is independent of m.

Theorem XI.

$$\frac{1}{2}\sum_{mm'} |(\alpha'I + 1m'|Y_{1\mu}|\alpha Im)|^2$$
$$= \tfrac{1}{3}(I + 1)(2I + 3)|(\alpha'I + 1I|U_z|\alpha II)|^2, \quad (5)$$

$$\frac{1}{2}\sum_{mm'} |(\alpha'Im'|Y_{1\mu}|\alpha Im)|^2 = \frac{(I + 1)(2I + 1)}{3I}|(\alpha'II|U_z|\alpha II)|^2, \quad (6)$$

$$\frac{1}{2}\sum_{mm'} |(\alpha'I - 1m'|Y_{1\mu}|\alpha Im)|^2$$
$$= \tfrac{1}{3}I(2I + 1)|(\alpha'I - 1I - 1|U_z|\alpha II - 1)|^2. \quad (7)$$

Equation (6) excludes $I = 0$. The proof is based on the explicit representation of the component matrix elements and the relation

$$\sum_{m = -I}^{m = +I} m^2 = \tfrac{1}{3}I(I + 1)(2I + 1). \quad (8)$$

5–7 Charge and supermultiplets. An interesting and important application of the angular momentum formalism occurs in the theory of nuclear charge multiplets. The theory is based on the formal possibility of treating the neutron and proton as different states of a fundamental particle, the nucleon. Two functions

$$a(\tau) = \delta_1(\tau) = 1, \quad \tau = 1$$
$$= 0, \quad \tau = -1,$$
$$b(\tau) = \delta_{-1}(\tau) = 0, \quad \tau = 1$$
$$= 1, \quad \tau = -1 \quad (1)$$

describe states in which the nucleon is definitely a neutron (a) or definitely a proton (b). Operators τ_1, τ_2, τ_3 are defined by the relations

$$\tau_1 a = b, \quad \tau_2 a = ib, \quad \tau_3 a = a,$$
$$\tau_1 b = a, \quad \tau_2 b = -ia, \quad \tau_3 b = -b, \quad (2)$$

in exact analogy with the Pauli spin operators $\sigma_x, \sigma_y, \sigma_z$ acting on spin functions $\alpha(m_s)$ and $\beta(m_s)$ describing states having definite values of S_z. The exact analogy is continued by the introduction of total charge spin operators

$$T_1 = \frac{1}{2}\sum_1^A \tau_{1k}, \quad T_2 = \frac{1}{2}\sum_1^A \tau_{2k}, \quad T_3 = \frac{1}{2}\sum_1^A \tau_{3k} \quad (3)$$

having properties formally identical with those of the total spin operators S_x, S_y, S_z. In the physical applications, the wave functions $\psi(\ldots r_k, m_{sk}, \tau_k \ldots)$ describe states with definite values of N and Z; hence T_3 has a definite numerical value fixed by the eigenvalue equation

$$T_3\psi = T_3'\psi, \quad T_3' = \tfrac{1}{2}(N - Z). \tag{4}$$

A useful starting point for the discussion of nuclear structure is the assumption of a charge-independent Hamiltonian (reserving for later study all charge-dependent effects). The Hamiltonian H then commutes with T_1, T_2, T_3, and $\mathbf{T}^2$; consequently, there exist simultaneous eigenfunctions of H, $\mathbf{I}^2$, I_z, $\mathbf{T}^2$, and T_3. The eigenvalue equation

$$\mathbf{T}^2\psi(\ldots |EImTT_3') = T(T + 1)\psi(\ldots |EImTT_3') \tag{5}$$

defines a quantum number T; solutions exist for $T_3' = T, T - 1, \ldots,$ $-T + 1, -T$. Thus corresponding states exist in a range of isobars extending from $\tfrac{1}{2}(N - Z) = T$ to $\tfrac{1}{2}(N - Z) = -T$; this set of corresponding states characterized by definite values of EIT and parity constitutes a charge multiplet. No analysis is required to establish these results, since they are immediately evident from the exact correspondence between the charge variable formalism and the theory of the intrinsic spin.

If both spin and charge-dependent interactions are neglected, two or more charge multiplets may coincide in energy, forming a supermultiplet. In this approximation L is a good quantum number common to all the coincident charge multiplets. The properties of supermultiplets are most easily established by considering the set of operators $\mathbf{T}$, $\mathbf{S}$, and

$$Y_{1x} = \frac{1}{2} \sum \tau_{1k}\sigma_{xk},$$
$$\vdots$$
$$Y_{3z} = \frac{1}{2} \sum \tau_{3k}\sigma_{zk}. \tag{6}$$

These operators commute with the spin and charge-independent Hamiltonian and with the parity and orbital angular momentum operators. The reader can verify that T_3, S_z, and Y_{3z} constitute a complete set of commuting operators in the sense that no others in the set of fifteen commute with all three.

Three numbers P, P', P'' may be used to characterize a super-multiplet: P is the largest eigenvalue of T_3 compatible with the given values of A, E, L, and parity; P' is the largest eigenvalue of S_z when $T'_3 = P$; and P'' is the largest eigenvalue of Y_{3z} when $T_3 = P$ and $S'_z = P'$. The wave function $\psi_{PP'P''; T_3'S_z'Y_{3z}'}$ belongs to the super-multiplet $PP'P''$ and is a simultaneous eigenfunction of T_3, S_z, and Y_{3z}, with the indicated eigenvalues. In principle, the complete function space associated with the supermultiplet can be generated by applying suitable displacement operators constructed from T, S, and Y's in succession to $\psi_{PP'P''; PP'P''}$. Operators with the property of displacing two of the eigenvalues independently upward or downward by one unit while leaving the third unchanged are particularly useful; the construction of such operators is left as an exercise for the reader.

To illustrate these ideas, we evaluate the Fermi and Gamow-Teller type nuclear matrix elements in the theory of allowed beta transitions. The Fermi operator, $T_1 \pm iT_2$, commutes with $\mathbf{T}^2$, hence has non-vanishing matrix elements only within a charge multiplet. Recognition of $T_1 \pm T_2i$ as a displacement operator yields immediately

$$(\alpha, T, T_3' | T_1 - iT_2 | \alpha, T, T_3' + 1) = [(T - T_3')(T + T_3' + 1)]^{\frac{1}{2}}. \quad (7)$$

In particular, for image transitions

$$(\alpha, \tfrac{1}{2}, \mp \tfrac{1}{2} | T_1 \mp iT_2 | \alpha, \tfrac{1}{2}, \pm \tfrac{1}{2}) = 1, \quad (8)$$

and for the $0 \to 0$ transitions at $A = 10$ and 14

$$(\alpha, 1, 0 | T_1 - iT_2 | \alpha, 1, 1) = 2^{\frac{1}{2}}. \quad (9)$$

The symbol α represents all quantum numbers other than T common to the initial and final states.

The Gamow-Teller operator,

$$\mathbf{M}^\pm = Y_{1x} \pm iY_{2x}, \ Y_{1y} \pm iY_{2y}, \ Y_{1z} \pm iY_{2z} \quad (10)$$

has nonvanishing matrix elements only within a supermultiplet. To evaluate these matrix elements for image transitions, consider the commutation relations

$$(T_1 - iT_2)(Y_{1u} + iY_{2u}) - (Y_{1u} + iY_{2u})(T_1 - iT_2)$$
$$= -2Y_{3u}, \quad (u = x, y, z). \quad (11)$$
$$T_3(Y_{1u} + iY_{2u}) - (Y_{1u} + iY_{2u})T_3 = Y_{1u} + iY_{2u}. \quad (12)$$

The second of these implies

$$(Y_{1u} + iY_{2u})\psi_{PP'P'';\,PS_{z'}Y_{3z'}} = 0, \tag{13}$$

since otherwise the left member of Eq. (13) would be an eigenfunction of T_3 with the eigenvalue $P + 1$. The first relation then yields

$$(Y_{1u} + iY_{2u})(T_1 - iT_2)\psi_{PP'P'';\,PS_{z'}Y_{3z'}} = 2Y_{3u}\psi_{PP'P'';\,PS_{z'}Y_{3z'}}. \tag{14}$$

In image transitions, $S = T = \tfrac{1}{2}$ and $PP'P'' = \tfrac{1}{2},\tfrac{1}{2},\pm\tfrac{1}{2}$. The charge multiplet notation $\phi_{SS_{z'};\,TT_{3'}}$ may be substituted for $\psi_{PP'P'';\,PS_{z'}Y_{3z'}}$ in Eq. (14), with the result

$$(Y_{1u} + iY_{2u})\phi_{\frac{1}{2},m_s;\,\frac{1}{2},-\frac{1}{2}} = 2Y_{3u}\phi_{\frac{1}{2},m_s;\,\frac{1}{2},\frac{1}{2}}, \tag{15}$$

from which follows

$$
\begin{aligned}
(\alpha,\tfrac{1}{2},m_s';\,\tfrac{1}{2},\tfrac{1}{2}|\,&Y_{1u} + iY_{2u}|\alpha,\tfrac{1}{2},m_s;\,\tfrac{1}{2},-\tfrac{1}{2}) \\
&= 2(\alpha,\tfrac{1}{2},m_s;\,\tfrac{1}{2},\tfrac{1}{2}|\,Y_{3u}|\alpha,\tfrac{1}{2},m_s;\,\tfrac{1}{2},\tfrac{1}{2}) \\
&= \pm 2(\tfrac{1}{2}m_s'|S_u|\tfrac{1}{2}m_s).
\end{aligned}
\tag{16}
$$

In Eq. (16) the second form of the right member follows from Theorem IV of Section 4–1. The proportionality constant relating the matrix elements of Y_{1x}, Y_{2y}, Y_{3z} to those of S_x, S_y, S_z is determined to fit $u = z$ and $m_s' = \tfrac{1}{2}$. In the I,m representation, Eq. (16) becomes

$$
\begin{aligned}
(\alpha,I,m';\,\tfrac{1}{2},\tfrac{1}{2}|\mathbf{M}^+|\alpha,I,m;\,\tfrac{1}{2},-\tfrac{1}{2}) \\
= \pm 2(\alpha,I,m';\,\tfrac{1}{2},\tfrac{1}{2}|\mathbf{S}|\alpha,I,m;\,\tfrac{1}{2},\tfrac{1}{2}) \\
= \pm 2(Im'|\mathbf{I}|Im)\,\frac{(\alpha II;\,\tfrac{1}{2},\tfrac{1}{2}|\mathbf{I}\cdot\mathbf{S}|\alpha II;\,\tfrac{1}{2},\tfrac{1}{2})}{I(I+1)}.
\end{aligned}
\tag{17}
$$

Consequently,

$$\sum_{m_f} |(f|\mathbf{M}^+|i)|^2 = \frac{4}{I(I+1)}\,\overline{\mathbf{I}\cdot\mathbf{S}}^2. \tag{18}$$

Here i and f refer to initial and final states and $\overline{\mathbf{I}\cdot\mathbf{S}}$ is the diagonal matrix element or mean value of $\mathbf{I}\cdot\mathbf{S}$ for either initial or final state.

REFERENCES

1. H. B. G. CASIMIR, *Interaction between Atomic Nuclei and Electrons.* Teyler's Stichtung, *Archives* (3) 8; 201–287 (1936).

2. E. U. CONDON and G. H. SHORTLEY, *The Theory of Atomic Spectra.* New York: Macmillan (1935).

3. P. A. M. DIRAC, *The Principles of Quantum Mechanics.* London: Oxford (1935).

4. E. C. KEMBLE, *The Fundamental Principles of Quantum Mechanics.* New York: McGraw-Hill (1937).

5. G. RACAH, "Theory of Complex Spectra":
   (a) *Phys. Rev.* **61,** 186 (1942),
   (b) *Phys. Rev.* **62,** 438 (1942),
   (c) *Phys. Rev.* **63,** 367 (1943).

6. E. P. WIGNER, (a) *Gruppentheorie und ihre anwendung auf die quantenmechanik der atomspectren.* Ann Arbor: Edwards Brothers (1943),
   (b) *Phys. Rev.* **51,** 106 (1937),
   (c) *Phys. Rev.* **51,** 948 (1937),
   (d) *Phys. Rev.* **56,** 519 (1939).

7. M. BORN and P. JORDAN, *Elementare Quantenmechanik.* Leipzig: Springer (1930).

8. C. ECKART, "Group Theory and Quantum Dynamics." *Rev. Mod. Phys.* **2,** 305 (1930).

9. L. I. SCHIFF, *Quantum Mechanics.* 2nd ed. New York: McGraw-Hill (1955).

10. M. E. ROSE, *Elementary Theory of Angular Momentum.* New York: Wiley (1957).

11. A. R. EDMONDS, *Angular Momentum in Quantum Mechanics.* Princeton: Princeton University Press (1957).

INDEX

Adjoint, 3
Angular momentum operators, 10

Born, 55

Casimir, 55
Charge independent Hamiltonian, 52
Charge multiplet, 51
Complete set of functions, 16
Commutator, 2, 7
Complete set of commuting operators, 6
Compound systems, 22
Condon, 55

Degenerate eigenvalue, 4
Dirac, 7, 55
Displacement operators, 9, 20, 53
Double-valued functions, 28

Eckart, 55
Eigenvalues of orbital angular momentum, 15
Electric dipole moment, 15

Fermi operator, 53

Gamow-Teller operator, 53
Gyromagnetic ratios, 44

Harmonic oscillator, 7
Hermitian operators, 3

Image transition, 54
Infinitesimal transformation, 14
Invariance, under rotation, 16, 18
Isobars, 52

jj coupling, 49
Jordan, 55

Kemble, 55

Ladder of solutions, 9, 20
Landé g-factor, 33, 44, 48
Laplace's equation, 46
Linear operators, 1
LS coupling, 49

Magnetic moment, 15, 44

Nuclear magneton, 44

Orthonormal set, 4

Parity, 45
Pauli spin operators, 26
Precession, 33

Quadrupole moment, 45

Racah, 55
Rigid rotation, 15
Rotation of axes, 12

Scalar product, 3
Shell model, 49
Shortley, 55
Simultaneous eigenfunctions, 4, 20
Spin operators, 11
Supermultiplet, 52

Tensor operators, 38, 42
Torque, 17
Transition probability, 50

Unitary transformation, 5

Wigner, 55